Science As It Happens!

SCIENCE
As It Happens!

Family Activities with Children Ages 4 to 8

Jean Durgin Harlan, Ph.D.
Carolyn Good Quattrocchi

An Owl Book
Henry Holt and Company   New York

Henry Holt and Company, Inc.
Publishers since 1866
115 West 18th Street
New York, New York 10011

Henry Holt® is a registered
trademark of Henry Holt and Company, Inc.

Published in Canada by Fitzhenry & Whiteside Ltd.,
195 Allstate Parkway, Markham, Ontario L3R 4T8.

Library of Congress Cataloging-in-Publication Data
Harlan, Jean Durgin.
    Science as it happens! : family activities with children ages 4 to 8 /
Jean Durgin Harlan, Carolyn Good Quattrocchi.—1st. ed.
        p.   cm.
    "An Owl book."
    Includes bibliographical references.
    1. Physics—Study and teaching (Elementary)   2. Science—Study and
teaching (Elementary)   I. Quattrocchi, Carolyn Good.   II. Title.
QC30.H33 1994   93-46660
649'.68—dc20   CIP

ISBN 0-8050-3061-1

Henry Holt books are available for special promotions and
premiums. For details contact: Director, Special Markets.

First Edition—1994

Designed by Kate Nichols
Illustrations by Laura Hartman Maestro

Printed in the United States of America
All first editions are printed on acid-free paper.∞

1 3 5 7 9 10 8 6 4 2

For her careful guidance through the publication process, we thank our editor, Jo Ann Haun. For helping us learn through our lifetimes, we thank our parents, grandparents, children, and grandchildren.

Contents

In vain have you acquired knowledge if you have not imparted it to others.

—Midrash from Deuteronomy Rabbah, 500 A.D.

PART ONE

Getting Started

CHAPTER 1
Why Early Science?

For young children, learning about our world and how it works begins at home—and early. You can make science *real* and fascinating for your child. And yes, *you* can do it.

You can prepare your child to enjoy science in school—to want to learn more about the important things the two of you have already noticed, talked about, played with, and explored.

You don't have to convert your house into a science lab—it already is one. You just need to recognize how to use it that way. You don't have to squeeze formal science lessons into your tight schedule. Informal experiences are more effective. It won't take a lot of time to turn your child on to science. You can do it casually as you go about your usual chores.

For example, the bathtub provides a natural place to learn about water. Why do some things float and others sink to the bottom of the tub? A walk together in the sunshine can reveal light and shadow; all of us—adults and children alike—find our own shadows fascinating.

You don't have to be a rocket scientist, or even know how your CD player operates, to help your child discover some of the powerful ways the physical world works. Some of the most basic ideas in physics and chemistry are not complicated.

Chances are that you are like most of us—a victim of dull

science classes that convinced you science is only for geniuses. You probably don't know how much you *do* know. And what you *don't* know is really not hard to understand, especially if you start at the beginner's level *with* your child.

So you *can* help your child find answers, calm fears, satisfy curiosity, and launch into school science eager and ready to learn more. You can make science a natural part of your child's life. You and your child can begin right now—together—to find out how our world works.

"But My Child Is Way Ahead of Me . . ."

You say your four-year-old is caught up in computer games that you don't understand? Your child knows how to program the VCR, but you don't? To clear up some confusion, you could have learned to use those electronic marvels just as quickly as your child did. But you had already learned a different way of doing similar things. Perhaps, for you, learning the new gadget involves overcoming learning resistance or slowing down to unlearn the old way first.

Knowing how to punch the right buttons to operate a device is a skill, like tying shoes. Having that skill doesn't mean that your child understands the basic science that makes electronic technology possible. Your skilled button pusher will still be intrigued by the simple ideas you can offer about static electricity—why, for example, when we walk across a carpet on a cold, dry day and then touch a dooknob, we get a small shock.

But you are right about your child's natural learning ability. You have seen the gifts of wonder, curiosity, and imagination that motivate your young child to want to learn. These marvelous qualities are inborn. You just need to support these natural gifts to help your child stay on the path to learning.

4

Overcoming Your Own Science Phobia

"So what happened to *my* natural curiosity about the world?" you may ask. "I never liked science in school, so how can I help my child stay interested in it?" Like countless others, you just need a bit of help first to break through the barrier of science phobia.

Removing a mental barrier begins with identifying its origins. Take a moment to recall the source of your science phobia. Was it a dreary class taught by someone who was required to teach science but was unable to teach it well? Was it a defeating course taught by a science whiz who couldn't connect with students? Was it an embarrassing moment in a science class that left you feeling inadequate? Were you afraid to take science electives because of their reputation for toughness? Did noxious aromas from a science lab turn you off from all science? Was it just not socially acceptable in your crowd to like physics or chemistry?

These, and more, are compelling reasons for a child or teen to dread science. But are they powerful enough to stop your adult self from being aware of how things happen around you? If not, then allow yourself a fresh start at home with your child. Join your child in becoming a careful observer, a fascinated explorer of simple events around you. The moment you recognize yourself sharing your child's satisfaction and delight in finding out, take note and congratulate yourself. You have successfully identified, confronted, and conquered science phobia!

When Does Science Teaching Begin?

Should you begin teaching the principles of gravity to a child in the playpen? Of course not. You don't have to. Infants are already busy making cause and effect discoveries about gravity—that toys fall to the floor when dropped from the high chair, for starters. But as toddlers begin to express themselves with words, you can supply them with accurate science words to label what is going on.

Children naturally grow up with beginning science words and concepts like *hot* and *cold*, *push* and *pull*, and *lift* and *move*. Because these concepts are not complex and perplexing, they may not seem to count as science. But they are genuine pieces of science information that children grow up with as accepted beliefs.

Some developmental theorists have argued that when children reach preschool age, they stop using the ability they had as infants to make cause and effect discoveries. But today's researchers are finding that preschoolers *can* use cause and effect logic, given the chance to explore and try out for themselves just "what happens when . . ."

You begin sharing your knowledge of how things work when your child's curiosity calls for a new word, a simple 10-second answer, or an easy new experience that can then invite new questions. Finding out together can become a comfortable, habitual way of gradually helping children absorb what they truly want to know.

Is Early Science Really a Family Affair?

Human beings have a natural urge to care for offspring, as they have done for thousands of years. That includes teaching them about the world they live in. For centuries parents have assumed this role. But public schools became necessary as family life changed with the Industrial Revolution. Before long, parents came to believe that only schools *could* teach their children.

In recent decades, educators have realized that the easiest children to teach in school are those who have already learned much at home. Today you, as parents, are encouraged to build the foundations of reading as soon as you can hold both a book and a baby on your lap at the same time. You are urged to talk about the numbers that stand for the amount of things your child can see or touch.

The older, narrow vision of formal, school-based education as the best way to teach children is changing. But, for the most part, the physical sciences are still deemed "too difficult for amateurs to teach." These are the very branches of science in which today's American high school students have been faring so badly.

But impressive evidence backs the idea that you can help your child with beginning steps toward future science learning. Benjamin Bloom, in his book *Developing Talent in Young People*, traces the course of talent development in 120 outstanding adults from their early childhood onward. The study found clear patterns of family support for children's curiosity in the early life of 50 science researchers and mathematicians.

Bloom reports, "In these homes the parents encouraged the curiosity of their children at an early age, and answered their questions with great care. . . . They were patient in helping the children learn very simple skills in these fields." Bloom found that the children saw the parents' support as recreation and play, *not as demands to learn.* Nor did the parents deliberately attempt to mold their children into future scientists.

Einstein spoke of the wonder he experienced about the age of five, when his father showed him a simple magnetic compass. "I can still remember . . . that this experience made a deep and lasting impression on me. Something deeply hidden had to be behind things." Einstein had only begun to talk shortly before that time, incidentally.

Beginning physical science is only "too difficult for amateurs to teach" when taken out of the context of the child's known world. It is only too difficult for children to understand when stripped of the appreciation, practical value, and delight that are possible when families explore together at home.

Children who come to school confident that curiosity leads to answers are the ones who bring enthusiasm to science lessons. They are eager to learn more. That's what families can do.

CHAPTER 2
How? New Approaches

Launching Points

Start with Your Child's Interests

Let your child's current interests and questions lead into your first family science venture. Beginning a project that sparks only *your* interests risks disappointment. Steve learned this the afternoon he tried to share his favorite technology museum exhibits with his son, who was not quite three years old. The only displays worthy of newly toilet-trained Andrew's attention were the impressive rows of plumbing in each men's room he explored.

A highly verbal child's questions about puzzling things are easy leads to follow up with a suggestion of, "Let's do something to find out more after dinner." Rachel was fascinated by her uncle's explanation of the bellows he used at the fireplace to start a brisk fire. Months later, she discovered an opposite effect when she blew out the five candles on her birthday cake. Pausing for a bite of cake, she asked, "Why does air put out the candle fire if you need air to start a fire?" That conversation stopper eventually led to a postparty discovery activity (see Fire Facts, page 179).

Less verbal children who learn best in more active ways may return repeatedly to intriguing problems. Their interests may come to your attention less readily, but once you recognize the interest, you can offer a related discovery activity.

Your Interests as Springboards

Is your youngster at your elbow watching as you take apart your camera for lens cleaning, or as you adjust your ten-speed for an outing? Such moments are natural opportunities for finding out more about lenses and light, or how gears help bike riders. Your connection with the shared discovery activity makes it more significant for your child. Also, your child benefits from seeing how you stay with an interest long enough to get good at it—a bonus for learning.

Daily Events as Learning Leads

A spontaneous household happening can stimulate a shared experiment to find out why it occurred. Your child might notice the crackle and spark as you pull a pair of jeans and a clinging sock from the dryer. Or you might call attention to the few rice grains quivering against the inside of the plastic bag you've just emptied. It doesn't matter who initiates the noticing and wondering why. What matters is your interest in helping your child develop an inquiring attitude. You do this when you respond to the interesting situation as something worth investigating. When you say something like, "I wonder what could be causing this to happen. Let's see what we can find out about it later," you are modeling the value of curiosity as a cause-and-effect thinking and learning tool. Learning from real life has a vitality and relevance that no school workbook lesson can match!

Fears as Learning Motivators

Scary sounds, spooky shadows, awesome lightning, or the in-evitable dark can lend urgency to finding out their causes. But understanding *why* certain things happen can loosen the grip of fear by giving children a degree of control over their worrisome thoughts.

Emotionally powered interests during childhood can have a long-lasting effect on learning directions. Matt's early fear was that a falling star could destroy his house during the night. To be certain that all the stars were still in place, he tried to count those he could see from his window before he gave in to sleep. Then he learned through many experiences that Earth's gravity is far too weak to pull down distant stars. The impact of that old fear, and the relief his new knowledge brought, kept Matt fascinated with astronomy throughout his school years. Now he is doing graduate work in meteorology.

Not all children are as open about the anxieties beneath their fascination with a subject. But it just might help a reticent child if you mentioned your own early worries that could have been eased, if only you had found out the cause of the frightening event.

Safety as Learning Reinforcement

Safety warnings can fade in a moment of childish fun. Taking the time to do some simple experiments together can strengthen those warnings with impressive, tangible meaning and prevent painful or even life-threatening accidents. Activities with safety implications for children are starred for you in Chapter 3.

Successful Sharing

Hands-On Plus

Children retain what they learn about science when they are actively involved in discovery. Hands-on learning offers several ways to absorb impressions, using all the senses and handling real materials. The actual doing provides added data to crystallize the learning. So, as much as possible, stand back and allow your child to do the pouring, lifting, holding, squeezing, and such. If you have any reservations about how all this might work out, try the activity when your child isn't with you. You can also adapt activity suggestions to fit your own circumstances.

Learning can be more engaging when children have a few playful minutes to get acquainted with materials you'll be using before you begin your experiment. Then, after the activity, allow time for them to confirm what they have learned by talking it over and by repeating the activity until they are satisfied they have mastered it. In the future reinforce that learning by pointing out new instances wherever you encounter them together.

Lab Equipment

The activities outlined in Part II of this book do not call for expensive equipment, nor do they require elaborate preparations beforehand. They are carried out with familiar, inexpensive recyclables, toys, or ordinary household equipment. Because your free time is limited, and because it isn't much fun for children to wait and watch while you put devices together before beginning the experiment, the activities are limited to those that use at-hand materials and are readily accomplished by young children. When they are old enough to read and follow complicated instructions independently, they will enjoy constructing their own equipment for experiments.

Patience, Please

It's great to share what you know. But sharing discovery activities is not about showing children how smart you are. It's about encouraging them to observe, try out, and figure things out for themselves as much as possible. As you work together, ask, "What might happen if . . . ?" and "Do you think it would work differently if we . . . ?" Then allow enough time for your child to think and search out answers. Better reasoning will result. Encourage trying out guesses and hunches, as professional scientists do. But don't leave matters up in the air if the guesses and results are way off target. Offer better alternatives. "Hmm, that was worth trying, but it might work better if . . ." Eventually you might need to share what you know about *why* and *how*. Each activity description in this book includes simple information about underlying principles to help you do that.

Your Feelings Count

Consider postponing a new science activity if you are feeling rushed or upset about something. When you explore with your child, your own emotional state makes an impact on what is learned. Like the rest of us, your child thinks and remembers best in a relaxed, unpressured mood. Actually, certain brain functions shut down when we are stressed, anxious, or angry.

Your Child's Feelings Count

By adult standards, young children can come up with amusing ideas to explain how or why things work. According to Christopher, the water stays in the straw when he holds his finger over one end and lifts it out of a glass of water, because "if it didn't stay there, it would make a mess, and that would be wrong to do." But please hold the laughter. Being laughed at is always demeaning. It is a discouraging reminder of how little a child knows, compared to grown-ups, who surely know everything.

Christopher's idea is a sign of an active mind and deserves respect. Actually, all of us form our own private theories to explain many things, filling in the gaps in our information with what we hope is true. At this point in Christopher's life, moral issues of right and wrong are very important, so he fit them into his conclusion about the water in the straw. Correcting misconceptions of this sort needn't be a big concern. Eventually, when more accurate information is both available and acceptable to the child, these misconceptions yield to understanding. Children are likely to hang on to private theories much longer when they fear being ridiculed or criticized for exposing their ideas.

Persistence Pays

It's not too hard to decide when a discovery activity has lasted long enough. It's easy to tell whether a child is engrossed or bored with experimenting. But it may not be as easy to recognize whether we have "overtalked" about it. Children know how to tune us out imperceptibly if, when they asked what time it is, we elaborate on how to make a watch. Yet, we know that superficial exposure to a new idea doesn't lead to durable learning.

New ideas need to be put to use many times before they move from short-term to long-term memory. The good news from learning research is that repetitions and reuse need not be routine drills. Instead, information is more indelibly retained when many different pathways of learning are used.

Casual reference to new information can be fun. For example, you could ask at mealtime, "Have you noticed that your crackers are floating on top of your soup? Are the carrots floating on top, too?" That's more intriguing than a drab, resistible suggestion like, "Now, let's talk about what you learned in that floating and sinking experiment last week." With a light touch, new ideas can fit into your family conversations in unlimited ways on various occasions.

Another important way to reinforce learning is to create new images with the facts. You and your child could play a game en route to the supermarket to visualize an idea in a new way. You might suggest, "Pretend I'm a Martian on my first visit to Earth. I've never seen water before. Can you tell me something that water does?" Or, you could read aloud a story that refers to something you have explored together; each activity chapter lists some related children's books. Or you might enjoy creating a jingle or a song about the learning with your child. Writers of advertisements know how potent these memory ticklers can be.

Whenever it's possible, as you travel with your family, include visits to the wonderful children's discovery areas in city science museums. Locate and use the science books in your children's library. Watch public television science programs for children and the science segments on *Sesame Street*. Before long, these kinds of informal support for science learning will flow regularly and naturally into family life.

Using This Book

Chapter 3 is designed as a chart to help you make the connections between everyday events and the science activities that reveal the science principles governing them. Usually it will be helpful to do more than one related activity within a topic area. Randomly choosing single activities from a variety of topics is less effective.

Each chapter in Part II represents a single science topic, with activities arranged from the simplest to the more complex. The As It Happens anecdotes opening each activity use "your child" in a general way to represent real questions that children have asked. Your own child may not think about and question events in the same way. Activity directions offer simple responses that youngsters of this age can absorb. Basic science principles are set out in **boldfaced italic** type. They can be shared verbatim with your child, but don't expect these principles to be fully understood. That takes time and experience. Certain activity descriptions include more detailed explanations for the benefit of the guiding adults. These explanations are set apart in brackets like these: [. . .].

In general, topics can be explored as interest and events dictate. However, *Weather* concepts are easier to grasp after some foundation experiences with *Air* and *Water*; *Gravity* activities make better sense when they follow *Magnetism* experiences; and *Static Electricity* activities make *Current Electricity* concepts more understandable. No other sequencing of topics is necessary. Learning from life doesn't unfold as a formal lesson plan. Over time your child's discoveries circle back and weave through everyday occurrences, enlarging and gaining meaning with each encounter.

As you share these family activities, you may discover a potential dividend for yourself. These reminders of the simple, yet powerful, laws of nature can renew and deepen your appreciation and enjoyment of life, making your efforts twice as worthwhile.

> *If a child is to keep alive his inborn sense of wonder, he needs the companionship of at least one adult who can share it . . . rediscovering with him the joy, excitement, and mystery of the world we live in.*
>
> —Rachel Carson, *The Sense of Wonder*

CHAPTER 3

When? As It Happens!

Countless occurrences in family life are natural curiosity starters. These are times for noticing, and later exploring, physical science as it happens. Here are some of the everyday settings, the common events, and the science activities that lead to satisfying "why and how" answers.

Don't push to cover all of the possible learnings at all possible learning opportunities. What matters is enjoying some of these activities at times that work best for both you and your child. These are only some of the ways to ease your child into appreciating and understanding the workings of physical science. You will find other means and moments to share this way of looking at the world with your child.

*Starred activities have the added value of explaining child safety rules.

As It Happens	Curiosity Starter	Discovery Activity	Page
In the Kitchen	Recycling plastic bags	*Bag It!*	29
	Basting foods	*A Pressing Matter*	35
	Cooking rice/dried beans	*Soaking It Up*	47
	Mixing beverage crystals	*Demystifying Dissolving*	50
	Making toast	*The Whys of Wind*	62

There are two ways to live your life. One is as though nothing is a miracle. The other is as though everything is a miracle.

—Albert Einstein

PART TWO

Getting Involved

CHAPTER 4
Air

Air Is Real Stuff

BAG IT!

As It Happens

Your child wonders "how come" punching down on a tiny plastic bellows makes an air pressure–activated toy zoom across the floor. You look at the bellows together to figure out what it pushes into the toy to make it go. Since children rarely notice invisible, ever-present air, this is a good beginning point for finding out what this vital substance can do.

Gather

2 Ziploc plastic bags

Help Find Out

Start things off by cupping your folded hands, with the thumbs side by side. Using wrist action, squeeze your hands together rhythmically to pump air out between your thumbs. Let it puff on your child's cheek. Ask, "Do you feel something? Do you *see* anything coming out of my hands? If you can't see something, can it be real stuff?"

Then ask, "Is the chair you're sitting on real? How can you tell? If you closed your eyes, how could you tell the chair is real? Is feeling something a good way to tell if it is real?" Now have your child take one plastic bag by an open edge and whirl it around to catch some air in it. Seal the bag quickly. "What did you catch in the bag? Look at it. Feel it. Do you think it is something real?" **We know that it is real stuff because it fills up space.**

Put the air-filled bag on one chair seat and the uninflated bag on another. Have your child sit on each seat. Do both bags feel the same? Why not?

This experiment, as well as other air experiments, might have been done with balloons instead of clear plastic bags. But the more transparent the container for the air, the more apparent is the concept that *invisible air* is taking up space inside the bag. We can feel air inside the bag, we can see the bag enlarge, but we cannot see the air.

Follow Up

Another time at the kitchen counter, let your child try to roll up emptied plastic bags from the open end to the closed end. What gets trapped inside? How can the bags be rolled up for recycling so they can be closely packed?

TUB TIME

As It Happens

Your child is squeezing air from a baster to push a toy sailboat *over* the bath water. You ask, "Do you think air can push on something *under* the water?"

Gather

Tub filled with bath water
Cork
Clear plastic tumbler

Help Find Out

Push the cork to the bottom of the tub. "The cork is touching the bottom now. What will happen if we let go of it? Let's see." Let your child push the cork down several times and let go of it.

Then ask, "Can you put the cork back on the bottom without touching it? Let's see if something in this tumbler will push the water away so the cork can stay on the bottom." Show your child how to invert the tumbler over the floating cork and push straight down to the bottom of the tub, taking the cork with it.

"What could be inside the glass pushing the water away? Do you see anything?" Let your child experiment, tipping the tumbler to let air out and water in. Watch the air bubble surface. *"We can't see air, but we can see what it does to other things."*

Follow Up

If you receive a package that uses bubble-wrap packing material, think about why this helps keep things inside the package from breaking. Then have fun popping the bubbles!

If you have a bicycle tire pump, fireplace bellows, or bellows toys, look at the air intake hole where air comes in to be pushed out by the pump. Let your child feel the air being pushed out.

If you have a bellows-type step pump for inflating such things as air mattresses, show your child the rubber cap on the air intake valve. Push the air out so the bellows are flat. Then remove the cap so that air rushes in and watch the bellows quickly expand.

BOOKS FOR YOUR CHILD: With his lighthearted style, Richard Scarry illustrates concepts about the air we breathe, moving air and its uses, and principles of flight, among others, in his *Great Big Air Book*, New York: Random House, 1971.

*Take a Deep Breath

As It Happens

Your child wants to zip-seal a caterpillar into a small bag to show at school. You intervene and explain that the caterpillar always needs a fresh air supply to live. "People do too, and here's a way to find out why."

Gather

Facial tissues

Help Find Out

Say, "Take a deep breath, shut your lips together, and pinch your nose shut for as long as you can." When the gasp for air comes, say, "What happened? Why did you let go of your nose? What did your body need that made you let go?" Try this a couple of times.

Then explain, "You needed to get a breath of air. All living things need fresh air to stay alive. Our bodies need the part of air that is called *oxygen*. Oxygen is used up so quickly inside us that we need to breathe in fresh air about every six seconds. That's why you must *never* try to hide or fit into a tight place where fresh air can't get in.

"Let's hold these tissues in front of our faces. See what happens to the tissue when you bring air into your body and then when you push this air out. Try it with your lips closed. Now open your lips a little bit. Can you feel the cool air coming into your nose? Into your throat?" Explore where the air goes once it is inside our bodies. "Look at your chest moving up and down. The fresh air coming into your body has oxygen in it. The used air that comes out doesn't."

BOOKS FOR YOUR CHILD: Franklyn Branley's *Oxygen Keeps You Alive*, New York: Thomas Y. Crowell, 1971.

---

*Starred activities have the added value of explaining child safety rules.

Air Is Everywhere

As It Happens

Your child is in the bathtub, playing with empty shampoo bottles or tub toys. It's an easy time to add a bit more information about ever-present air.

Gather

Paper or Styrofoam cup
Bathtub or bathroom sink filled with water
Empty clear plastic bottles, uncapped
Floating tub toys

Help Find Out

First, look together at the cup. "Is there anything in it?" Then let your child punch a hole in the bottom. "If something is in the cup, could we push it out through this hole? Let's see." Invert the cup and slowly push it straight down into the water. Have your child hold a hand just above the hole. "Do you feel something being pushed out of the hole?" (A stream of air?) "Could something be real in that empty cup . . . something real that we can't see?" Let your child experiment.

Now examine the plastic bottles. "Do you see anything inside? Point the bottle top toward your cheek and squeeze it. Do you feel something? What? Try to squeeze it out."

Suggest that you try to aim the bottle at a floating tub toy and squeeze. "If your bottle is really empty, nothing will come out to push the toy around. See what happens."

After much bottle squeezing has occurred, ask, "Is a container, even when it looks empty, ever *really* empty?" **No, air is almost everywhere.**

Follow Up

"Catch some air in your hands. Now poke your nose into your hands. Can you smell the air? Peek in. Can you see it? Can you hear it? Can you taste it?" Let your child explain to *you* how we can know that air is almost everywhere and that it's real. This time, your child can be the expert, while you play the skeptic.

Play a game together, thinking of all the funny places air can be: in a mouse's ear; in the holes in Swiss cheese; inside a keyhole, etc. "We are surrounded by air. We live in air, just as fish live in water."

When you make beds with your child, explain why you shake the pillows and comforter: "We're letting more air get inside. The bits of down or fibers trap tiny pockets of air to make pillows plump and soft and comforters warmer to sleep under."

BOOKS FOR YOUR CHILD: *Whistle for Willie*, by Ezra Jack Keats, New York: Viking Press, 1964. Use this story as an example of safe play with a box. We need to have enough fresh air to breathe wherever we play.

Air Presses Up, Down, and Sideways

A PRESSING MATTER

As It Happens
You may need to divert a restless child in a casual restaurant. You could offer to pick up some water from your glass without touching it.

Gather
Drinking straw
Glass of water
Clear plastic kitchen baster (use later at home)

Help Find Out
Let your child look inside the straw. Is it empty? Now insert the straw in water. Put your index fingers over the top and bottom of the straw to lift it from the water. Keep your finger on top of the straw and remove the other finger from the bottom. Most of the water will stay in the straw. Why? Because air pressing up under the straw keeps it in—*air presses on everything on all sides.* Say, "Nothing can get into the straw from the top with my finger on it. Watch as I let some air push into the straw." Remove your finger from the top and watch as the water is pushed out of the straw. The greater amount of air above the straw can now press down into the straw to push the water out. Let your child experiment until the meal is served!

At home, extend your child's understanding of this idea by using a clear plastic kitchen baster. Again, let your child examine the baster first to see that it looks empty. Then squeeze the bulb to feel what comes out. Put the baster in a jar of water. "What do you see when you squeeze air out of the baster? Watch the tube when you slowly let go of the rubber bulb on top. What is happening?" Lift the baster out of the water. Squeeze the bulb to push out the water. Release the bulb to pull in air. "What's happening now?"

Follow Up

Notice toothpaste oozing out of the tube after your child has stopped squeezing it: air pressure at work again. Notice that catsup flows out of the bottle easily after you let some air slip inside to push the sauce out.

HOLDING POWER

As It Happens

At bath time your child is experimenting again to see how a tumbler full of air can push water away. You offer to explore what happens in reverse: when air pushes on a tumbler full of water.

Gather

Clear plastic tumbler
Plastic container lid to fit the tumbler
Bathroom plunger
Facial tissue

Help Find Out

Fill the tumbler brimming full of water. Press the lid over the top of the tumbler to form a tight seal between the rim and the lid. Holding the lid firmly in place, invert the tumbler. Carefully let go of the lid. (The lid stays on. The water can't spill out.) Together, tip the covered tumbler various ways to be sure that air presses on the lid from all directions. [It does. The pressure of the air outside the tumbler is greater

than the pressure of the water inside the tumbler.] *"Air is always pressing on things from all sides.* Air always presses on us, too, but we are so used to it that we don't notice it pressing on our bodies." Be prepared for an extended bath time while your child enjoys doing this activity.

Release the lid to empty the water into the tub. Let your bath-time scientist continue to experiment with the new discovery.

Offer another experience with air pressure: After the bath, put the plunger on the floor. (A smooth surface works better than a small-tile floor.) Let your child tear up small bits of tissue and place them around the rim of the suction cup on the floor. Watch as one of you forces the plunger down hard. What happens to the tissue? What pushes it away? (Air from the rubber cup.) What happens to the suction cup? (It collapses a bit.) What do you hear? Can your child lift the plunger off the floor now? (No, the force of the air outside the cup is much greater than the pressure of the air left inside the cup. The cup is held fast to the floor.)

Follow Up

Look for suction cups holding signs in store windows, decorations on car or home windows, gadgets in your kitchen, such as towel racks and hooks that attach with suction cups. You'll find several beneath a rubber tub-safety mat.

EASING ON DOWN

As It Happens

Your child is gazing out the window on a fall day and sees leaves drifting down very slowly from a tree. You might observe that there isn't much wind to blow the leaves down fast. Then you could suggest making something to figure out what kept the leaves from falling down faster today.

Gather

2 pieces of typing paper

Man's handkerchief or a 12-inch square of thin fabric

4 pieces of light string or thread, each a foot long

2 small toy action figures or unbreakable rings from the toy shelf or 2 rubber washers

Help Find Out

Have your child scrunch one piece of paper into a tight ball. Hold the ball in one hand, the sheet of paper horizontally in the other hand. "If I hold these high and let go, do you think they will fall the same way?" Find out. "Can you think of anything beneath the sheet of paper that slowed its fall?" (More air pushed up under the wide sheet of paper than under the small ball, so the sheet fell slower.) *"Everything that moves above the ground must push air aside as it moves."*

Now, with your other materials, make a parachute. Help your child tie one piece of string to each corner of the handkerchief. Then knot the strings together as shown and fasten the ends of the string to the toy figure, being sure that the figure weighs more than the parachute.

Fold the parachute fabric and wind the strings and chute around the toy figure. If you have a stairway, let your child drop the toy without an attached parachute and then the toy with the parachute, from the top of the stairway through the banisters. Which one reaches the floor first? Outdoors, have your child throw the parachute as high as possible into the air and watch it open as it falls.

BOOKS FOR YOUR CHILD: *Georgie and the Runaway Balloon,* by Robert Bright, Garden City, NY: Doubleday, 1983. Georgie turns a handkerchief into a parachute to rescue a mouse with a runaway balloon. *Curious George Gets a Medal,* by H. A. Rey, shows George floating back to earth from a space flight with the help of a parachute. New York: Scholastic Book Services, 1957. In both these stories, the air beneath the parachutes allows them to drift slowly to earth.

Uplifting Action

As It Happens

During a drive to the airport on a very windy day, your child may worry, "What if the wind pushes Grandma's plane to the wrong airport?" You reply that big jet engines and special wings help keep the plane moving in the right direction. Moving air helps lift and keep the plane up. Later your child can make a glider to find out how moving air pushes a glider along.

Gather

Sheet of newspaper
Sheet of typing paper
Ruler
Pencil
Paper clip

Prepare

Draw a line lengthwise down the center of the typing paper.
Draw parallel lines 1½ inches from the center line.
Draw a line ¾ inch from the bottom across the paper.

Help Find Out

Give your child the sheet of newspaper. "Hold the paper against you. Will it stick to you if you let it go while you're standing still?" (No.) "Now do that again, but this time run across the room or down the hall. What's different this time? What could be holding the paper against you?" (Air moving against the moving child.) *"Moving air pushes against things.* It's the same whether we are doing the moving, or the *air* is moving and pushing against us as *wind* when we

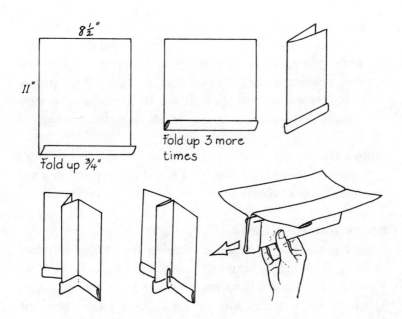

Fold up ¾"

Fold up 3 more times

are standing still. Let's make something fun for moving air to push along."

Show your child how to make a small glider out of the piece of typing paper. If you have a favorite way of making a paper airplane, by all means use it. If not, here is an easy construction method your child can use.

First, fold up ¾ of an inch at the bottom of the paper. Fold this over again three more times. With this cuff on the outside, fold the paper and the cuff along the lengthwise center line. Now fold back each side on the lines 1½ inches from the center. This makes the wings. Fasten the front fold with the paper clip as shown.

Launch the glider with the cuff in front, holding the fold. If the glider always does a nosedive, reduce the weight on the front end by unfolding one fold of the cuff. Also, try to launch the glider with the arm thrusting straight ahead. Reinforce the idea that air moving under the glider wings lifts and supports it and lets it drift ahead as long as the force of the push lasts. Fast-moving air beneath their wings helps lift the planes and keep them up, like the paper gliders. But planes also have special wings and jet engines or propellers to keep them moving through the air (thrust) in the right direction, no matter which way winds might be blowing.

Follow Up

Look for an old-fashioned balsa wood toy airplane in a toy store. It has a propeller powered by a twisted rubber band.

BOOKS FOR YOUR CHILD: *Paper Flight*, by Jack Botermans, offers directions of increasing complexity for making 48 paper gliders and 15 origami birds and insects. New York: Henry Holt, 1984. Older children will appreciate hearing you read *Flight*, by Robert Burleigh, an outstanding book about Lindbergh's first flight across the Atlantic. New York: Philomel Books, 1991.

RESOURCES: The Nature Company offers a Flight Kit of experiments to move objects through the air. Designed by the Boston Museum of Science, the kit includes directions to make and fly gliders, kites, and make a parachute and helicopter twirler. Send for a catalog by calling 1-800-227-1114.

CHAPTER 5

Water

Heavy Stuff

As It Happens

It's time to fill up the wading pool on a hot day. You suggest moving the pool into the shade before you fill it with water. Otherwise, the pool will be too heavy to move. Your child protests, "But water isn't heavy!"

Gather

Large bucket

Hose

Help Find Out

First, shake a few drops of water onto your child's hand. "Does this feel heavy? Do you think water *could* be heavy? Let's find out."

Let your child hold the bucket under the hose as you fill it with water, or if indoors, under the bathtub faucet. "Now what do you think?" ***Water has weight.*** "The deeper the water in your bucket, the heavier the bucket becomes."

Follow Up

Learning about water comes naturally when you visit the swimming pool or beach with your child. Try to run in knee-deep water together. It's hard work! You are pushing away water's weight as you run.

*Will It Float?

As It Happens

At bath time your child stuffs the wet washcloth into a floating toy boat to keep the cloth handy. The plan fails. Both sink to the bottom of the tub. You help find out why.

Gather

Assorted objects: rocks, sticks, bath toys, corks, metal washers. Try to find pairs of objects that look alike but are made of different materials—such as a golf ball and a Ping-Pong ball.

Balloon

Heavy aluminum foil

Wooden cutting board

Hammer

2 clear plastic shampoo bottles with tops

Help Find Out

First, just let your child play with the things you've assembled. "Does the water hold some things up on top? All the things? Why do you think this ball stays on top and that ball sinks down?"

Have your child hold the Ping-Pong ball in one hand and the golf ball in the other. Which is heavier? Then explain, "The weight of the water pushes up under things. **Water's weight helps things float.** The golf ball is too heavy for the water under it to hold it up, but the Ping-Pong ball is light enough to stay on top."

Next, blow up the balloon. Let your child push it to the bottom of the tub. Let go. Watch how fast the water pushes the air-filled balloon to the top.

Form a small, flat-bottom boat from the aluminum foil.

Watch it float on the water. "Let's see if the foil will float if it changes shape." Crush the foil into a ball as tightly as possible. Put it on the wooden cutting board, pound it into a compact lump with a few hammer blows. Does it float now or sink? (It sinks.) [There was more water under the long boat shape to hold it up. The spread-out foil was less dense than the condensed lump.]

Remove the top from one of the empty shampoo bottles. Submerge the bottle and let it fill almost to the top with water. Recap the bottle. "What is inside the bottle now? (Air and water.) Is the air or the water on top? Which will be on top if we turn the bottle upside down? On its side? Find out." [Air is lighter than water, so it's always on top.]

Add the second capped bottle, this one full of air. What happens? "The air inside this bottle is helping it float. *Air can help things float.*"

Follow Up

At the swimming pool or beach, let your child compare the difference between curling up in a ball in the water and stretching out to float on top of the water. What happens? (You will probably have to help younger children accomplish this.)

Then let your child stand shoulder deep in water. "Spread your arms away from your side just a bit and *feel* the water push your arms up to the top." It takes effort to push them back down against the weight of the water.

BOOKS FOR YOUR CHILD: In Marjorie Flack's classic book *A Story About Ping*, a young houseboy is kept afloat, after tumbling into the Yangtze River, by a barrel of air tied to his back. New York: Viking Press, 1970. And in H. A. Rey's *Curious George Rides a Bike*, George forgets to deliver his newspapers because he is so busy making newspaper boats to float downstream. New York: Scholastic Book Services, 1973.

Soaking It Up

As It Happens

Your child has just shed a wet swimsuit after playing in the wading pool or coming from the beach. Your reminder that wet suits shouldn't be left on couches or beds may raise the question, "Why not?" Here's a fast way to find out.

Gather

Hard, dry sponge
Baster or medicine dropper
Water
Materials to test, such as small piece of wood, cotton ball, paper towels, stones, graham cracker, dry cereal
Piece of dry cloth
Spray bottle
Small piece of plastic sheeting or rubberized fabric
Rubber boot

Help Find Out

Put a wet suit on a paper towel briefly. Let your child check the towel to discover that water can go into certain things. It can even move from one thing to another.

Put the hard sponge on a dry place. Let your child fill the baster or medicine dropper with water, then squeeze water onto the dry sponge. "Watch carefully. Something will change fast!" (A dry sponge absorbs water and expands with surprising speed.) Explain the term *absorb*. Explore what happens to drops of water squeezed from the baster or medicine dropper onto other test materials. Which one does water go into (absorption)? Which ones do not let water in? [Water molecules move into tiny spaces and cling to absorbent materials.]

For the next experiments, drape the piece of dry cloth over your child's arm. Spray with water. "What do you feel?" Repeat with the plastic sheet over your child's arm. "Now what do you feel? Which material would you want for a raincoat? Why?" Let your child wear a rainboot on one foot and have the other foot clad only in a sock. Spray both feet with water. Which foot has a covering that is *absorbing* water? Which does not?

Follow Up

If you can find a stiff bird feather, let your child discover how birds stay dry in the rain. Sprinkle a few drops of water on it. What happens? (The drops aren't absorbed. These stiff outer feathers have an oily coating to shed water.)

The next time you cook rice, let your child measure the dry rice before adding it to the boiling water and after the rice is cooked. "What do you think happened to the water?" Do the same with dried beans before and after an overnight soaking in water. Measure the soaked beans before and after cooking them. Save a few dried beans to compare next day with the cooked beans.

If you have a small kitchen scale, your child can verify that water can go into certain materials by weighing a sponge when it's dry, then when it's soaking wet. To double check what made the difference in weight, squeeze out the water as much as possible and weigh the sponge again.

From now on, use the term *absorb* when your child needs to wipe a spill. It just may soften the drudgery of clean-up!

For an absorbing boredom breaker during a casual

restaurant wait, pinch off one end of a paper drinking straw cover. Slide the cover down and crush it into a tightly pleated "caterpillar." Slip it off the straw. Lift a few drops of water with the straw (see page 55) and drip them on the center of the crushed cover. Watch the tube move around as the paper absorbs the water and expands.

Demystifying Dissolving

As It Happens

Your child is stirring soft drink crystals into a pitcher of cold water and observes, "I'm doing magic. I made the powder disappear! I'm turning water into lemonade!" This is a good time to let your child discover what dissolving is about.

Gather

Muffin pan or plastic ice cube tray with separated cube molds

Water

Assorted dry materials: salt, sand, cornstarch, flour, sugar, cornmeal, etc.

Taster spoons or other small spoons for stirring

Small screw-top bottle

Salad oil

Help Find Out

Let your child half-fill the individual muffin cups with water, perhaps using a baster. "See what happens when you put a little salt in one of your muffin cups of water. Stir it. Can you see it? Feel it? Where is it?

"Now, dip your finger in the cup. How does the water taste? The salt is still there, but it is in such tiny bits that it can't be seen. We say that the salt *dissolved* in the water when each little salt grain broke into tinier bits in the water." [*Dissolve* means "to break apart." Water molecules break apart the salt molecules. Salt and water molecules join to form a solution.]

Let your child experiment with stirring the other dry materials in separate pans of water to see which ones will dissolve. ***Some things dissolve in water and some do not.***

Half-fill the bottle with water. Add some oil. Cap the bottle securely and let your child shake it. Did the oil seem to dissolve? (Perhaps.) Let the mixture stand for a while, then look again. Where is the oil now? Did the oil *really* ever dissolve in the water? (No.)

Follow Up

The next time you make Jell-O together, you will *both* know what happened to the powdered Jell-O. You can reinforce this concept with other dissolving foodstuffs: sugar, instant coffee or tea crystals, or bouillon crystals, for example.

A Dry Subject

As It Happens

Perhaps your child promises to finish a half-eaten sandwich later, but when "later" comes, the sandwich is disappointingly dry. Your child wonders why.

Gather

Paper towels
Water
2 trays
Hand-held hair dryer
Small blackboard
Letter-size sheet of cardboard
2 pie pans
Plastic wrap

Help Find Out

Together, wet some paper towels. Spread them out on the two trays. "What do you think will happen if we put one tray in a sunny or warm place and the other in a cool, darker place? Let's find out." Check them in half an hour.

Now turn on the hair dryer and feel the warm air coming out of it. "This warm air can help us understand what will happen to the towels."*

Let your child dip a finger in water and draw or write with it on the blackboard. "Now blow some warm air from the hair dryer on your wet blackboard picture. Watch what happens to the water. Where did it go? Only the air touched it. The water has gone into the air in such tiny drops [vapor] that we can't see it anymore. When air picks up water this way, we say the water *evaporates*."

---

*This is a good time to remind your child that we *never* turn on the dryer near a tub or sink filled with water.

Let your child draw another picture with water on the blackboard. "Now take this piece of cardboard and wave it near your picture to make a breeze. Which evaporates the water faster, hot air or cooler air?"

Check the two paper towels again to see if the water evaporated faster in the sun-warmed air than in the cooler, shady air.

Compare what happens to a tablespoon of water left overnight in an uncovered pie pan and the same amount of water left in a pie pan tightly covered with plastic wrap. Droplets of water couldn't leave the covered pan. They could leave the uncovered pan. (Did some water droplets collect under the plastic? See Cycling Water on page 64 for more condensation ideas.)

Follow Up

Compare fresh with dried fruit: grapes and raisins, plums and prunes, and fresh and dried apricots. What brought about the change?

Point out other examples when they occur to reinforce the notion that moisture in food left uncovered is picked up by the air (the food dries out). Compare a piece of bread left uncovered with one sealed in a sandwich bag. See what happens to food left for a while in the refrigerator without a cover. The refrigerator is full of air, too!

Think about what happens to wet mittens when they are stuffed into coat pockets, rather than being spread out to dry in the air. And wet bathing suits left wadded up in a beach bag stay just that way—good and wet. Compare drying times for two damp towels after bath time: one tightly rolled up, the other well spread out on the towel rack. (Now your child has the rationale for a good housekeeping habit. It may help keep things tidy!)

Let your child feel the moist warm air that rushes out when the clothes dryer is opened mid-cycle. The warm air is pulling moisture from the damp clothes.

Finally, you can make holiday cookie ornaments, feeling the dampness of the dough while you're mixing it. Save a small lump of damp dough in a plastic bag.

HOLIDAY "COOKIE" ORNAMENTS

This recipe is for crafts only; not for eating.

1 cup corn starch
2 cups baking soda
1½ cups water

In a large pot, combine the above items. Stir until smooth. Cover and cook over medium heat stirring occasionally, until mixture is the consistency of slightly dry mashed potatoes. This will take about 5–6 minutes. Turn onto a cool surface and cover with a damp cloth. When cool enough to handle, coat hands with corn starch and knead until dough is smooth and pliable. Roll out and cut with holiday cookie cutters. Use a toothpick to make holes for hanging. To dry, place on a baking sheet in a 250-degree oven until hard. Ornaments may be painted after drying.

Compare the damp dough to the baked ornaments. "Where did the dampness go?"

Water's Surface Is Tight

PULLING TOGETHER

As It Happens

During a long, nervous wait at the clinic, your child gets many drinks at the water fountain. You offer a more interesting diversion: water stretching.

Gather

Water

Waxed paper or smooth countertop

Small medicine dropper or straw to pick up and dispense
water

Clear plastic tumbler

Baster

Uncooked spaghetti

Help Find Out

While you are still in that waiting situation, show your child this new way of looking at water. Put a single drop of water on your child's thumb. "Touch the water drop *very gently* with your finger. Watch carefully." When your child's forefinger touches the drop on the thumb, the water drop sticks to both. Pulling the thumb and finger apart gently will stretch the drop of water; pushing them together compresses it. "We'll find out more later about stretchy, sticky water drops."

At home, give your child a square of waxed paper. "What will happen if a tiny drop of water is squeezed carefully from the dropper (or straw) onto the paper? Find out. Did it make a flat splash or a rounded drop?"

Then see what happens when many drops are made

close together. (The drops pool and flatten out; the curved top is gone.) Talk about how the outside edges of the drops pull together to make an invisible "skin." This "skin," called *surface tension,* is not very strong.

Next, fill the plastic tumbler almost to the top with water. While your child watches closely, slowly add drops of water from the baster until the jar is full to the brim. Ask, "What might happen if more drops are added slowly?" Find out, drop by drop. (The surface pulls more drops into a curved dome of invisible "skin" until it can hold no more.)

Fill the tumbler again. Hold a piece of spaghetti with your thumb and forefinger. Carefully rest it horizontally on the surface of the water. (It stays there!) *"Surface tension can hold some weight."* Now push it through the "skin" vertically. This time, it sinks. (It broke through the surface tension.)

Follow Up

Cover a mug of very hot beverage with a clear, plastic yogurt cup lid. Use a magnifier to examine the beautiful, perfectly curved, condensed water droplets that collect under the lid.

After a rain, notice perfectly curved drops of water everywhere: on leaves, spiderwebs, or the surface of a polished car.

Use water's surface tension to make wonderful soap bubbles. Adding detergent to water increases the stretchiness of the water. Use about ½ cup of good liquid dish-washing detergent for each quart of water.

Try using a fat drinking straw as a pipe for blowing small bubbles. Show your child how to dip the end of the straw into the solution and let a film collect across the bottom of the straw. (Hold the straw slightly downward to avoid dripping soap into the mouth.)

Use a plastic funnel to blow giant bubbles. First, gently blow a few bubbles into the bowl to allow a film of solution to coat the inside of the funnel. Lift out the funnel and blow softly.

Be sure to talk about how the outside edges of the soap film pull together like a balloon around the air. The soap mixture makes a more flexible "skin" than the water alone can make.

BOOKS FOR YOUR CHILD: *Bubbles,* by Bernie Zabrowsky, offers directions for making imaginative bubbles. Boston: Little, Brown, 1979.

RESOURCES: Children's science museums often have fascinating giant bubble-making demonstrations, perhaps even pulling a soap film around a child who stands in a solution-filled pool. Most good toy stores and children's museum shops carry giant bubble kits.

BREAKING THROUGH

As It Happens

Seeking a shortcut at hand-washing time, your child may complain, "Why do I have to use *soap?*"

Gather

Bathroom sink filled with water
Pepper or talcum powder
Liquid soap
Sliver of bar soap
Baby oil or petroleum jelly

Help Find Out

Fill the bathroom sink with water. Sprinkle pepper or talcum on the water. (Pepper shows up better in a white sink;

talcum shows up better in a colored sink.) "Does the pepper stay on top of the water or sink? What could it be resting on?" (Surface tightness or tension.) "The surface of the water stretches tightly like the outside of a bubble. It keeps those bits from sinking down."

"Watch closely now." Drop *one* drop of liquid soap in the center of the pepper. What happened? Refill the bowl and sprinkle again with pepper. Touch the center of the floating pepper with the sliver of soap. Is the action the same? (Yes.)

Explain that **soaps break the surface tension of the water.** The water surface no longer sticks to itself.

Next, pour some baby oil or spread some petroleum jelly on the palm of your child's hand. Sprinkle drops of water on it. "Tip your hand. What happens?" (The drops roll off.) "Are the drops sticking to the oil?" (No.) [Oil molecules stick together; water molecules stick together. They can't stick to one another: they can't mix.] Now drop liquid soap onto your child's palm. "Rub your hands together under the running water. What happens to the oil?" [Soap and water molecules *can* join together to break and keep apart, or *emulsify*, the oil molecules. The oil can then be washed away.]

"Dirt sticks to oil that comes from our skin, so *that's* why you have to use soap!"

BOOKS FOR YOUR CHILD: *Molecules*, by Janice VanCleve. Written for older children, this book offers a detailed explanation of surface tension, as well as further experiments. New York: John Wiley & Sons, 1993.

CHAPTER 6

Weather

Sun Sense

As It Happens

You arrive at the beach on a sunny day. Your barefoot child steps onto the hot sand, then dashes to cool off those hot feet at the water's edge. You hear, "The water is too cold, and the sand is too hot. Why can't they both be just right?"*

Help Find Out

You explain that ***the sun's energy warms the Earth and everything on it, but some things warm faster than others.***

Suggest digging a deep hole in the sand to find out more about how the sun warms the sand. What feels different as the hole gets deeper? Walk out into the water together, noticing changes in the temperature as the water becomes deeper. You may find a depth where the surface water is markedly warmer than the water below. If permissible, collect a small bucket of sand to continue experimenting at home.

Gather

2 shallow dishes or pie pans
Sand or garden soil

*This could also occur with the cement or tile apron at the swimming pool. Try to find a flat rock or brick to turn over, noting temperature differences between the top, warmed by the sun, and the cooler underside.

Outdoor thermometer*
Notebook
Pen

Help Find Out

Fill one container with water, one with sand. Check and record the temperature of each. Place both containers in a sunny location for 30 minutes. Check and record the temperatures again. Which one changed the most? Repeat on a cloudy day.

Another time, check and record the temperatures this way for two containers of sand and soil: one in the sun, one in the shade. And another time, check and record the temperature of one container of sand in early morning, noon, and after sundown.

Hang up the thermometer outdoors to check the air temperature with your child frequently. Let your child know that **the sun's warmth, together with air and water, make the weather.**

Follow Up

When you take a walk together, touch a sidewalk, building, car, or rock that is in sunlight. Then touch one in shade. Comment again that the sun gives the Earth both warmth and light. When the sun's warmth and light are blocked, it is cooler and darker. "Will it be as warm tonight as it is now in the sunlight?"

Watch the weather news with your child to verify that the air temperature is cooler at night and warmer in the day, when it is warmed by the sun. **The sun never stops shining. We just can't see it at night or on days when clouds cover the sky.**

---

*It is possible to feel the difference by touch, if you lack a thermometer.

BOOKS FOR YOUR CHILD: *The Sun, Our Nearest Star,* by Franklyn Branley. New York: Harper & Row, 1988.

RESOURCES: Hobby stores and good toy stores carry inexpensive packets of sun-sensitive paper for making fascinating prints on a sunny day.

The Whys of Wind

As It Happens

High winds are keeping everyone indoors. Your worried child watching at the window asks, "What makes the wind blow so hard?" You can offer to help find out what wind is.

Gather

Tissue wrapping paper
Scotch tape
Thread

Help Find Out

Say, ***Wind is moving air, and air is always moving.*** Let's find out if the air inside is moving." Tape a small strip of tissue paper to one end of a piece of thread. Tape the other end of the thread to a door frame, so that the paper dangles freely. Watch as the paper drifts on the air currents.

Next, explore the ups and downs of air movement. Carefully let your child hold a hand just above a lighted 100-watt light bulb to feel the warm air. Then, hold a ½-inch strip of tissue paper horizontally above the light bulb.* Let go of one end and watch the free end of the paper strip flutter up as the warm air from the glowing bulb pushes against it. ***Warm air rises up.***

Next, let your child hold the strip of paper just below the refrigerator or freezer door as you open it a crack. The cold air pushes the paper end down. ***Heavier cool air goes down below warm air.***

Talk about how wind happens: ***All over the world some air is being warmed by the sun, and some air is***

---

*Remind your child never to place paper directly on a light bulb. It will burn.

cool—over oceans, lakes, and other places. The cool air moves down under the warm air as the warm air rises. The spinning earth moves air along, too. All that pushing and rushing air is what we call wind.

Follow Up

When you are outdoors on a cold, calm day with your child, notice smoke rising up from the chimneys. Notice clouds of steamy air flowing up from clothes dryer vents. Notice steamy breath flowing upward from our mouths and noses.

On a very cold day, try blowing a few soap bubbles outdoors. Do they float up or down? (Up. Warm breath inside the bubbles rises over the cold air.) Compare this bubble path with that of bubbles blown indoors in warmed air.

Show your child how to find the wind direction on a calm day by holding up a wet finger.

Hang a wind sock in an open spot near your house. Check it for wind direction regularly. [The wind is named for the direction from which it blows. A north wind blows into the open end and pushes the wind sock toward the south.]

Enjoy the wind! Hang wind chimes or streamers in the yard. Fly a kite. Show your child how to launch the kite by running into the wind. Kites rise *against* the wind, not *with* it—a good analogy for facing life's challenges.

BOOKS FOR YOUR CHILD: *What Makes the Wind?* by Laurence Santrey. This simple, clear book is illustrated with watercolors of breezes and gales that almost lift off the pages. Mahway, NJ: Troll, 1985. *Feel the Wind,* by Arthur Dorros, includes simple directions for making a weathervane. New York: Thomas Y. Crowell, 1989.

Cycling Water

As It Happens

The TV weatherperson reports that clouds will be moving in to bring afternoon rain. Your child wonders, "But how do clouds get the rain?" You offer some ways to find out.

VISIBLE VAPOR

Gather

Glass or metal cup
Water
Ice cubes

Help Find Out

"There is always some water in the air. Air picks up bits of water from everywhere (see page 53). We can only see it when air is very full of water bits [vapor] close together as hot steam from a kettle or moist air in the bathroom after a hot shower. We see it as fog or clouds outdoors.

"Let's see what happens when warm air cools down next to something very cold." Let your child fill the glass with water and ice. Note that the outside of the cup is *dry*. Let it stand in a warm place for 15 minutes. "Has anything changed?" [There was water vapor in the warm air touching the cup. The vapor condensed into water droplets on the outside of the cold cup.] ***Warm air holds more water than cold air. Water bits left the air that cooled and collected on the outside of the cup as water drops.***

Follow Up

Open the dishwasher after the rinse cycle to let dishes air dry. Let your child hold a cool, dry tumbler in the steamy interior of the machine to collect droplets.

Notice water condensing on the cool bathroom mirror after a hot shower.

Slip a clear plastic container lid over a mug of steaming hot coffee. Lift off the lid in a few minutes, and together examine the beautiful tiny water droplets that have condensed under the lid.

RAINMAKING

Gather
Ice cubes
Small foil pan (pie pan)
Very hot water
Clear plastic tumbler (dishwasher safe)
Flashlight

Help Find Out
Let your child heap ice cubes into the small foil pan. Pour about an inch of very hot water into the tumbler. Put the pan of ice on top of the tumbler

"We'll pretend this is like the outdoors. The air under the pan is cold, like the air high in the sky. [Cold air moves

down.] Pretend the water in the bottom of the tumbler is like a puddle on the ground."

"What do you see happening?" Shine the flashlight into the rising water vapor and the cloud it forms under the pan. After a few minutes, briefly lift the pan to see the tiny droplets condensing on the bottom of the pan. Replace it.

Keep watching together until the tiny droplets collect to become larger, heavier drops that eventually fall back down into the water, like raindrops falling from a cloud.

Explain that air near the Earth is warmed by the sun. It picks up water from puddles, lakes, leaves, dewy grass, and whatever moisture it touches. Invisible water in the air condenses into droplets as clouds when air is cooled high in the sky.

Cold droplets form drops big enough to be seen, and heavy enough to fall as rain, just as they did inside the tumbler. It happens that way to the same water over and over again. *The sun, water, and air make this cycle.* [To be accurate, condensation occurs in clouds when droplets collect on dust particles to form raindrops. You should say that the cloud in the tumbler is formed *almost like* the way rain is formed.]

*Freezing Facts

As It Happens

You and your child are walking outdoors together on a cold day. You come upon a partially frozen, shallow ice puddle. You wonder, "Why do you think the puddle is frozen around the edges, but not in the center?" When you get home, help find out.

Gather

Soup bowl or plastic storage bowl
2 identical, clear plastic screw-top bottles or safety-cap
   aspirin bottles
Plastic bag

Help Find Out

Fill the bowl with water, almost to the top. Put it in the freezer. Check it each hour. Which parts freeze first? [Ice crystals form first where cold air and cold containers touch the water: the outside freezes first, then the middle.] "Ice forms from the top downward." When you find a neat shell of ice with water in the middle, slip it out of the bowl to let your child play with it in the sink. Discuss these freezing patterns to teach a safety lesson if you live near a pond or lake that freezes in winter. In freezing weather, water outdoors cools and freezes more slowly than the air and soil around it. Ice crystals form faster and thicker around the edges of a pond or lake, then spread across the top toward the middle.

Next, let your child find out why we don't leave liquid-filled containers outside in freezing weather. Completely fill the two plastic bottles with water. Fasten the lids tightly. Leave one bottle out on the counter. Loosely fasten the plastic bag around the other bottle and put it in the freezer for several hours. Be careful when you take the frozen bottle out of the plastic bag, as there may be sharp bits of plastic within the bag. Compare the frozen results with the unfrozen bottle of water. "What happened?" [Water molecules spread out, or *expand*, as water freezes.] "The bottle broke as the water expanded and froze into ice."

Follow Up

Kids love to stamp through ice puddles. Now, when you walk together, investigate first to see freezing patterns within the puddles.

Check the top surfaces of ice cubes in a freezer tray to find the bump of expanded ice on each one.

Rainbow Making

As It Happens

Your child is awed by a lovely rainbow after a downpour and asks "Why do we have to wait for a rainy day to have rainbows?" You offer to help find out why.

Gather

A sunny location
Small pitcher of water
Shallow baking pan

Rainbow

Mirror

Water

Small mirror
Piece of white paper

Help Find Out

Explain that rainbows form when sunlight happens to shine just the right way into air that holds just the right amount of water after a rain.

Put the pan on a table next to a sunny window. Pour an inch or more of water into the pan. Slide the mirror into the water. Lean the mirror, facing the sun, against the edge of the pan. Have your child stand next to the table, back to the window and holding the paper so the mirror will reflect onto it. Adjust the angle of the mirror to reflect the sunlight bending [refracting] as it goes through the water. The rainbow colors appear!

Then let your child verify the part water plays in creating the rainbow effect. Place a hand over the water, just above the submerged end of the mirror. What happens? (The sunlight can't shine through the water to refract, so no colors appear. See page 146, The Secret in Light Beams, to find out how the spectrum separates into a rainbow.)

Follow Up

Standing with your backs to the sun, help your child see the rainbow colors reflected in water droplets spraying from a hose on a sunny day.

CHAPTER 7
Magnetism

Selective Services

As It Happens

Your child wants to expand the refrigerator door art gallery to include the cupboard door but, when the magnet refuses to cling to the wooden cabinet, decides the magnet has "stopped working." Help your child explore which materials are attracted by magnets.

Gather

Small magnet from hardware store*
Basket or bag filled with:
> Steel items—key chains, paper clips, safety pins, bolts, nails
> Nonsteel odds and ends—corks, coins, plastic spoons, stones, toys
Two empty plastic jars

---

*An inexpensive toy magnet may be too weak for the experiences in this chapter. While it actually takes a very strong magnet (about 15-pounds pulling power) to damage electronic equipment in the house, it's good practice to restrict magnet use near the computer and floppy disks, video or audio tapes, or other electronic equipment.

Help Find Out

Start with the magnet and paper clip on a table. Let your child slowly slide the magnet toward the paper clip. Then, to feel the effect of the magnet, put the paper clip in your child's palm.

While magnetic force can't be seen or felt, the *effects* of magnetism can be seen and felt in your child's hands. Children can accept the reality of an invisible force when they can have experience putting that force to work.

Next, ask, "Do you think the magnet will pull everything to it, the way it did the paper clip? Let's test these things in our bag."

After your child has experimented with the objects you've assembled, suggest sorting the objects into the two jars—one jar holding things the magnet will pull and the other jar holding things it won't. You or your child could label the jars YES and NO. "The things in this YES jar may not look alike, but they are all made out of *iron* or *steel*. **Magnets pull only on things made of iron or steel.**"

Follow Up

Let your child explore your house, hunting for other surfaces where a magnet could cling to hold a picture. Check the bathtub with a magnet. Older tubs are cast iron, covered with porcelain. Newer ones may be fiberglass. Are the plumbing fixtures made of steel?

Go on a magnet hunt. Look for hidden magnets throughout your house. You'll find unseen magnets that help us every day: magnets in paper clip holders, cupboard door catches, flashlight holders, message holders, handbag clasps. A magnetized rubber strip holds your refrigerator door closed. Show your child the magnetized strips on bank cards that you slide into automatic bank teller machines to activate them, on hotel doors and parking lot key cards. Let your child know that we *need* magnets and use them every day.

Finally, keep the YES and NO jars of sorted items. Your child can add things to the jars and explore again the effects of magnetism. This is a good kitchen activity to keep hands busy while you do chores.

RESOURCES: Try to include a magnet-activated game on your family toy shelves. The Nature Company catalog usually offers good magnet toys. The Nature Company Catalog, P.O. Box 2310, Berkeley, CA 94702. Others can be found on novelty store shelves.

Just Passing Through

As It Happens

Your child doesn't want magnets showing on a new picture for the refrigerator door gallery. The magnets stick to the door, but the picture won't stick on top of the magnets. Your child wonders why. You provide a few materials to help find out.

Gather

Steel thumb tacks, safety pins, paper clips
2 empty clear plastic jars or tumblers (use plastic for safe handling)
Styrofoam packaging tray
Sand or dirt
Water
Paper, cardboard, aluminum foil
Magnets*

Prepare

Put a steel object, such as a safety pin, in the tumbler. Put some steel objects in the tray and cover them with sand. Put water in another tumbler and drop a paper clip in it.

Help Find Out

First, check to see whether the magnet will attract paper, foil, and cardboard. "Now, let's see if a magnet can pick up a paper clip that is covered with a piece of paper." Try covering the clip with cardboard and aluminum, too. "Will mag-

---

*The thickness of the nonattracted material and the strength of the magnet will influence the success of this experiment. A strong, new magnet will attract a paper clip even through a fingertip. A weak magnet will not attract a paper clip through cardboard.

netism pull *through* these things to pick up steel?" (Effects will vary with the power of the magnet.)

Touch the magnet to the outside of the first tumbler. Does it attract the safety pin inside? Slide the magnet around the outside of the tumbler. What happens?

Next, dip the magnet into the sand-filled tray and the tumbler of water. *Magnetism passes through materials that it does not attract* to pull out the paper clips and pins.

Your child may find that some magnets in the house are stronger than others. Dropping or banging magnets weakens their pulling power. You can help keep horseshoe magnets strong by remembering to put the steel "keeper bar" that came with it across both ends.

Your child may be fooled by appearances and believe that the keeper bar is also a magnet—one that just doesn't work when separated from the magnet. To help understand the difference, substitute a stainless steel spoon for the keeper bar. "This spoon isn't a magnet, is it? Does the magnet attract it? Let's leave this spoon across the magnet while you experiment with the keeper bar to see if it works like a magnet." Remember, your child's own discoveries are much more convincing than any scientific statements you might make.

Follow Up

Make a "scrap-dance box." Have fun putting together a magnet toy. You'll need a small magnet of any type, a shallow box, some plastic wrap, tape, and a steel wool pad. Shred very small bits of steel wool from the scrubbing pad. Heavy kitchen shears work best for this job. Cut enough steel wool scraps to barely cover the bottom of the box. Stretch the plastic wrap over the top to make a cover and secure it to the box with tape. Let your child enjoy making the steel scraps dance and creating patterns by pulling the magnet beneath the box.

On a rainy day, your child can put on a puppet show with small wood or rubber figures from the toy shelf. Tape a paper clip to the bottom of each puppet. Put an open shoe box on its side to make a stage for the puppets. Your child can hold the magnet inside the box and use it to move the puppets around on stage.

You can even use magnetism to help with cleaning up. If you spill tacks or nails, let your child use a magnet to make quick work of cleaning them up.

Pushing and Pulling

As It Happens

Your child is having fun with a pair of magnets and wonders why sometimes they "grab together" and sometimes they push away.

Gather

String or thread

2 bar or horseshoe magnets

Nail polish or dot stickers (unless magnets are already marked north and south)

Pencil

2 ring magnets

Help Find Out

Tie one end of the string to the center of the magnet (balancing it horizontally if you are using a bar magnet). Hold the other end of the string in your hand.

Let the magnet swing freely. When it stops, one end of the bar (or one side of the horseshoe) magnet will be pointing north. Mark this end with the sticker or dot of nail polish. "This is the north-seeking end" [pole].

After you have marked the north-seeking ends of both magnets, let your child hold a magnet in each hand and try to touch the like ends (north to north or south to south) together. "What do you feel? Try to touch the unlike ends. What happens?" *The like ends of two magnets push away from each other (repel) and the unlike ends pull together (attract).*

Next, slide a pencil through ring magnets, with the like sides facing together. Watch them "float" apart. "Could an invisible force be pushing them apart?"

Give your child lots of time to explore this repelling and attracting force of magnets. With strong magnets, the results are fascinating. To be sure your magnets retain their strength, make a point of storing bar or ring magnets in pairs with opposite poles touching.

BOOKS FOR YOUR CHILD: *Experiment with Magnets,* by Helen Challand, offers a more complete explanation of the makeup of magnets and their effects; written for older children. Chicago: Children's Press, 1986.

End Power

HORSESHOE AND BAR MAGNETS

As It Happens

Your child may hold a horseshoe magnet by the two ends and try to use the curved middle to pick up a paper clip. "This magnet is broken. It won't work."

Gather

Horseshoe magnet

Paper clips

Steel key chain that is longer than the bar magnet, opened to full length

Bar magnet

Help Find Out

Your child is right about the weakness of the horseshoe magnet's curved middle. "Let's try it another way. Use the *ends* of the magnet to pick up the paper clips. Do the *ends* act differently from the middle?" (Horseshoe magnets are curved to put the ends closer together, so they have stronger pulling power.)

Next, touch the chain with both ends of the bar magnet. Notice whether the whole chain clings to the magnet or the middle part hangs free. ***A magnet's power is strongest at each end.***

Dangle the chain above the center of the bar magnet. The pull of a strong magnet will curve the end of the chain toward an end of the magnet.

RING AND DISC MAGNETS

As It Happens

After becoming comfortable with the notion that horseshoe and bar magnets have "end power," your child may be confused seeing a ring or disc magnet. "This one doesn't work the same way because it doesn't have ends. It only has sides."

Gather

2 carrots
Paring knife
Disc or ring magnet
3 paper clips

Help Find Out

To clear up the confusion, place two carrots parallel on a cutting board. "Each carrot has a north end and a south end." Cut midway across one carrot. Place the halves parallel to the whole carrot. "Now each of these pieces also has a north and a south end (side)." Next, cut a ¼-inch slice of carrot, stand it back on its edge, and point out the north and south sides of the slice. "Ring and disc magnets are made this way like slices of a long magnet. *The power of the magnet always collects at each end,* no matter if it is long or short."

Cut the whole carrot lengthwise to make some flexible strips. "Some magnets are bent in a horseshoe shape like this because the power is stronger when the ends are close together."

Check out just where on the disc or ring the magnet's strength is greatest by attracting the three paper clips. (More than three clips bunched together are too many for a clear demonstration.)

Follow Up

Magnetic materials are often mixed with rubber and ceramics. Look around your house for any items that have magnetic materials mixed in them, such as recording tapes and refrigerator door seals.

Make a Magnet

As It Happens
Clumps of paper clips dangle upside down from the paper-clip holder. Your child says, "Look, some of the clips have turned into magnets." You suggest an easy way to check that possibility.

Gather
Paper clips
Steel wool
Strong steel magnet (bar, horseshoe, or ring)

Help Find Out
Have your child separate a clip from the dangling bunch, while you tear off a few shreds of steel wool. "Can the paper clip attract the steel shreds?" (No.) "Did the clip become a magnet?" (No.) Explain that the magnet in the clip holder was powerful enough to *pull through* many paper clips. "It didn't change the clips into magnets, but you can make that happen."

Straighten out one clip. Show your child how to stroke it firmly across one end of the magnet, *always in the same direction*. Count aloud 25 or more strokes.

"Now try to attract the steel shreds with the clip." (It should be weakly magnetic.) **It has become a temporary magnet.**

This is a good time to show how jarring a magnet weakens it. Hit the magnetized clip against something hard a few times. Now try to pick up a pin or a bit of steel wool.

Follow Up
If refrigerator and cupboard doors with magnet closures are being banged shut, you now have an extra reason for requesting gentler closings: so the magnets will stay strong.

Make a Compass

As It Happens

You are on a family hike, using a compass. Your child asks, "How does a compass work? How does it know which way is north?"

Gather

Wine bottle cork (available in hardware or kitchenware stores)
Paring knife
Bar magnet
2-inch-long needle
Tape
Empty margarine tub
Water
Compass
Permanent marker

Prepare

Cut a thin slice (½ inch) off the end of the cork with the knife.

Help Find Out

Talk about your earlier experiences finding the north and south ends of magnets. ***Our Earth acts as though there were a giant magnet running through its middle from north to south.*** (The *North Pole* means the north end of the magnetic force.) ***The north-seeking end of a magnet always points toward the magnetic North Pole of the Earth.***

Let your child magnetize the needle as shown on page 82, holding it by the eye end. Tape the needle to the cork. Fill the empty margarine tub with water. Float the cork on the water. Watch what happens. Try to keep the cork floating freely until the needle stops.

Which way is the tip of the needle pointing? Try it several times to see that it always ends up pointing north. Mark the sides of the tub NORTH, SOUTH, EAST, and WEST. Describe how you figure out the other directions from the north-pointing needle. You now have a compass!

Check your floating compass with your regular compass. Hold the regular compass flat or put it down on the counter. Do both needles point north?

Follow Up

Put a bar magnet in an open plastic container large enough to hold it. Float the magnet in the bathtub at bath time. What happens? (If your tub is made of porcelain over cast iron, the magnet will be attracted to it.) Use a second magnet to chase the magnet "boat" around the water for fun. Which direction does the magnet point to when it stops moving, if your tub is not made of cast iron? (North.)

CHAPTER 8
Gravity

Gravity Pulls Us Down

As It Happens

Your child waits at the swing, complaining, "I get the next turn on the swing when it stops, but it's never going to stop!" You give assurance that an invisible force will make the swing stop . . . the same one that makes the slide and teeter-totter fun to use: gravity.

Later, find out: Can your child slide back *up* the slide? Stay up high on the teeter-totter without a partner? Then find out more about gravity at home.

THE INVISIBLE FORCE

Gather

Magnet
Paper clip
Child's chair

Help Find Out

The existence of the force of gravity is easier for children to grasp if they have already experienced the more tangible force of magnetism. Let your child watch a magnet's force pull a paper clip along a table. Say, "We can see what a magnet's pulling power *does*, but we can't *see* the power itself. That force is invisible."

Then make the comparison to gravity. ***There is also a giant, invisible force that pulls and holds down everything in our world. That force is called gravity. Gravity pulls on everything all the time.***

Now, have your child sit on the chair with both feet on the floor, back touching the chair back, and hands folded in the lap. "Try to stand up without moving any of your muscles or swaying your body forward. Try very hard." (It can't be done.) "What do you think held you to the chair when you tried to get up without moving forward?"

Explain that gravity made it impossible to straighten up without using muscle energy to work against the force of gravity. "Gravity kept you down on the chair, just as it keeps almost everything from floating off the earth. We are just so used to gravity that we don't even notice it pulling on us."

BOOKS FOR YOUR CHILD: *Gravity Is a Mystery,* by Franklyn Branley, doesn't explain precisely what gravity is because scientists haven't been able to describe it yet. New York: Harper and Row, 1986.

A WEIGHTY MATTER

As It Happens

Your child tests a new thought about gravity, "Does gravity only grab me when I'm sitting down or swinging? I don't feel anything pulling on me when I stand and walk." You agree that we don't usually notice gravity's pull. "Let's try to find a way."

Gather

Bathroom scale

2 full food cans of equal weight

Market bag with handles
Long mirror

Help Find Out

"One way we can find out how much gravity pulls on us is by standing on a scale. We call that amount our weight." Weigh your child on the bathroom scale. Then have your child hold a food can in each hand. Weigh again and compare the two weights. Which way is gravity's pull greater, with or without the weight of the cans?

Then have your child hold both cans of food in the market bag with one hand, so that the weight is all on one side. Weigh once more, to be sure that the amount of weight hasn't changed. "Now look in the mirror. Are you standing up straight or leaning over? You lean away from the heavy side to keep your balance.

"Now watch in the mirror as you change back to holding a can in each hand at your sides. Are you standing straighter? Now gravity pulls the same on both sides. Your body is balanced."

Follow Up

When you and your child are in a "waiting" situation or doing errands in the car together, play the imagination game, "What If . . . ?" What if the world had no gravity to hold things down? Think of all the funny things that might happen: You could look up in the sky and see a car floating by like a kite; you would try to eat your lunch only to find your plate drifting up to the ceiling. You'd be floating up in the air, too!

No Tipping, Please!

As It Happens

Your child is struggling to keep a tall, narrow-based block tower from toppling. You suggest starting again with longer, heavier blocks at the bottom. The wider base balances* the tower load, and it's a steady success!

Gather

Full-length mirror

Help Find Out

"From the time we begin to stand and walk, ***our bodies learn to balance against gravity's pull so we don't tip over.***"

Stand together in front of the longest mirror in the house to watch your changing reflections. First stand still and straight. Then lift one knee as high as possible. "Did you notice anything change a bit?" Stand on both feet again and try lifting the other knee, watching closely for a slight sway as weight shifts from both feet to one foot. "Can you feel something changing in your body when you do this slowly?"

Suggest, "With your arm down at your side, press one side of your body against the wall. Even your foot should touch the wall. Try as hard as you can to lift the other foot off the floor. What happens?" (You can't shift your weight to balance your body to do this. The wall is in the way. Your body would tip over if you lifted your foot.)

*"Balance" is an age-appropriate term, although this is actually a center of gravity concept.

Follow Up

Look for examples of figure skaters, dancers, and gymnasts as they spread their arms wide to balance on one leg. Comment that the performers worked hard to learn how to use their muscles and their energy to balance or push away from gravity's force.

The Ups and Downs of Balancing

As It Happens

Your child is at the park, perched up in the air on the teeter-totter, while a larger child on the other end is stuck on the ground. Your child asks why it isn't working right.

Gather

At the park:

Teeter-totter

At home:

Ruler

Half-circle block

Pennies

Help Find Out

While you are still at the park, help the children examine the teeter-totter closely. Look for the balance point [fulcrum] (the support under the middle of the board that holds it up against gravity's pull).

Have the children sit again on the two board ends. Let

the larger child move slowly toward the balance point or fulcrum until the board is brought into steady balance [equilibrium]. "Now is the pull of gravity the same on both sides of the teeter-totter? Now you can push up and down."

At home, experiment by balancing a ruler on the curve of a half-circle block. Then use stacks of pennies in various positions to balance unequal weights of pennies on the ruler. When the ruler is in balance, remind your child, "Now gravity is pulling the same on both sides."

Follow Up

If you can supply a toy balance for your child, also furnish a basket full of items to weigh and compare. Do things of similar size or appearance necessarily weigh the same?

RESOURCES: Toy balances for children are made by several manufacturers, one of which is Constructive Playthings. Their Rocker Scale is available by mail at 1227 E. 119th St., Grandview, MO 64030, or by telephone 1-800-832-0224.

Fun with Gravity

As It Happens

Your child wants to go outdoors to try to keep gravity from slowing down the swing, but bad weather cancels the experiment. You suggest some indoor fun *with* gravity.

Gather

2 small round balloons of the same color
¼ cup of sand
Small funnel
Small potato
2 table forks
2 heavy teaspoons

Help Find Out

Blow up one balloon and tie it. (Use balloons of the same color, so your child won't believe that color makes a difference in the results of your experiment.) Using the funnel, pour the sand into the second balloon; then blow it up and tie it. Put both balloons on a table or the kitchen counter and let your child try to blow them over the edge. Which one rolls off? Which one tips but stays on the edge? How does gravity help? If you have a toy balance scale, you can compare the weights of the two balloons.

Let your child try to balance the small potato on one finger. Then insert the two forks into opposite sides of the potato, with the handles slanting outward and downward. Gravity's pull on the forks is lower and farther out, so the potato now balances steadily.

Follow Up

Look for similar balancing toys that are made in India and some countries in Africa. These can be found in import gift shops.

Dealing with Doubts

Your child may wonder why gravity doesn't make clouds and helium balloons, for example, fall down. Explain that warm clouds and helium balloons are lighter than the air beneath them. Gravity *does* pull on the clouds and balloons, but it pulls *more* on the air. The air weighs more (gravity pulls more on it), so warm clouds and helium balloons float on the air.

Solar Security

As It Happens

Your child is awed by a spectacular sunset and wonders, "Where does the sun land when it comes down?" You can explain that the sun does not ever really go down and touch the Earth. "Our eyes fool us sometimes."

Gather

Child's small basketball or "earth ball"
String shopping bag or mesh produce bag
Strong cord, 5 or 6 feet long
2 ring magnets
Pencil

Prepare

Put the ball in the bag and tie the bag to one end of the cord.

Help Find Out

The sun is *very* far from the Earth. The sun has powerful gravity that *always* keeps the Earth the same distance away as it spins and circles the sun.

"Here's something like the way that happens." Slip the two ring magnets, with their *like sides together*, over the pencil. This is a tiny model of the distance apart being always the same. [Actually, the magnetic force is repelling the two like sides, keeping them "floating" the same distance apart, even as you move the pencil about.]

Now move outdoors where there is plenty of room for the next part of the experiment. Let your child pretend to be the sun. Have the ball represent the Earth. Holding the end of the cord, let your child swing the ball around in a big circle, turning around in place, to let the ball (or Earth) travel in orbit around the sun. "Do you feel the pull in your arm? That is somewhat like the sun's gravity pull on the Earth.

"The Earth always goes around the sun in the same elliptical path [orbit]. The sun's gravity keeps it there. The moon always circles the Earth in its same path [orbit]. The Earth's gravity keeps it there. But the stars are too far away for Earth to pull on them. They can never fall to Earth."

Follow Up

Notice TV reports of space capsule lift-offs. It takes a very strong force to push far away from the Earth's gravity pull. Once the space ship has pulled away, then it can also orbit the Earth. The astronauts inside float because they are too far from Earth's gravity to stay on the floor.

Here's an enjoyable way to remind your child that the Earth travels around the sun: Refer to your child's age as the number of trips he or she has made around the sun.

BOOKS FOR YOUR CHILD: Gravity is one of the topics covered in *Joe Kaufman's Big Book About Earth and Space,* by Joe Kaufman. New York: Golden Books, 1987. *To Space and Back,* by Sally Ride and Susan Okie, describes astronauts' experiences doing ordinary routines weightlessly in an orbiting space capsule. New York: Lothrop, 1986.

CHAPTER 9
Simple Machines

Need a Lift?

As It Happens

Your child wants to help unload luggage from the car, but the bags are too heavy to lift. You suggest using the luggage carrier to help. "It's a lever on wheels. Push *down* on the handle to lift the bag *up*, then roll it along behind you." Later you look more closely to see how a lever makes work easier.

Gather

Fold-up lugage carrier or grocery cart
Broom

Help Find Out

Examine the luggage carrier carefully. Notice that the wheels are the resting point [fulcrum] of the lever, the part that stays in place, so that the push down on the handle can lift the luggage. You can mention that a wheelbarrow is this kind of lever, too.

When you are next at the park or playground, you be the "load" to be lifted on the teeter-totter. Let your child provide the push (effort) to lift the load to discover that ***the closer*** *the load is to the fulcrum and the farther the push is away from the fulcrum, the easier the work of lifting the load.* You move closer to the resting point of the teeter-totter, while your child remains at the end. What happens if you reverse positions? (Your child can't possibly lift the parent/load.)

"When you used the board to lift me *up* by *pushing down*, the teeter-totter board became a lever. The board moves on this resting point. It must have a resting point to become a lever. Then the energy of your push *down* makes the other end of the lever lift up. The part that stays put to make this work is called a *fulcrum* or *resting point*. What would happen if you tried to use a board to lift a load without a fulcrum?" (The load would slide off.)

At home, when you are sweeping, introduce another kind of lever that helps with the work of pushing. "Show me how to get nice, steady sweeps with this broom. Notice where you have each hand: One hand holds down here, and one hand is close to the end of the broom. Now feel carefully what each arm is doing as you push the broom along." The arm on top pushes in the opposite direction that the bristles are sweeping. The lower arm hardly moves at all. It becomes the fulcrum: The broom works as a kind of lever. After your child has swept for a while, ask which arm gets tired from providing the effort, and which arm does not. "Try sweeping with your hands side by side. Does it work as well this way?"

Follow Up

The next time you and your child are working in the garden or clearing a path in the snow with a shovel, slow down long enough to identify that one arm is making the shovel a lever. Notice how much easier it is to lift snow or dirt this way than if your hands are side by side. Levers make our work easier!

At the grocery store, look for dolly carts or platform dollies being used to move cartons around the store or to unload trucks.

Perhaps the most impressive use of the lever principle can be demonstrated by jacking up a car! Just be sure to use all safety precautions, particularly that of blocking the wheels on the ground securely.

Easing Up

As It Happens

Your child may be struggling to lift a heavy box of blocks onto a low table. You suggest an easier way.

Gather

Folded ironing board (or a plank) at least 4 feet long (you also could use an indoor toddler slide)

Help Find Out

"Let's see if we can make it easier to lift your box of blocks onto the table." Lean the ironing board from the floor to the table. Place the box of blocks near the bottom of the ironing board.

"We call this a ramp. A ramp holds some of the weight of the box of blocks as you slide it up. Now you can move the box onto the table. *The ramp made lifting easier.*"

Follow Up

In public buildings, notice the ramps created for wheelchair accessibility. Point out the ramps at curbs that make it easier to lift bikes, wagons, or strollers from the street to the sidewalk.

Spiral Power

As It Happens

Your child, who has already discovered how a ramp makes lifting easier, is screwing down the lid of a jar. You ask, "Did you know that a twisted ramp makes the lid stay on? Let's find out how."

Gather

Crayon
Small recycled business envelope
Pencil
Screw
Screw-top plastic jar
Bolt with nut
Screw-type cookie press
Facial tissue
Old lipstick tube

Help Find Out

Draw a thick crayon line diagonally from one corner of the envelope to the other. Cut or fold along the line, leaving a crayon-edged triangle. "Does this line on this triangle remind you of a ramp?" Then wind the triangle around the pencil so that the crayon line looks like a screw thread. Let your child compare it to a screw. *A screw is a curved ramp. We use it to lift or push things.* Show your child the threading on a plastic jar and inside its screw-top lid. Then find out how those two screws can be threaded together so they can't be pulled apart. Now let your child screw the nut onto the bolt. "You can make the nut go up the ramp that curves around the bolt."

Take the cookie press apart and examine the pieces.

Remove the design plate and stuff a tissue in the press. Let your child see what happens when the screw knob is turned. "What lifted up the tissue?"

Pry apart the casing from the lining of the lipstick tube bottom to discover the variation of the screw principle that raises and lowers the lipstick.

Follow Up

Let your child experience being lifted up by the screw action on an old-fashioned piano stool or a swivel office chair if you have one available.

Children who live on or visit farms may see the screw

principle on the equipment used for harvesting crops or moving grain from bins to trucks.

Notice the space-saving—and fun—spiral slides in fast food restaurant playgrounds. They look like the wound-around ramps called screws.

If you have a home workshop, demonstrate how C-clamps and the vise work.

Let your child watch if you have occasion to use a corkscrew.

Friction Facts

As It Happens

Your child's hands are cold from being outdoors on a winter day. Or, perhaps in summer your child's hands and arms are chilled after a swim. "Rub the palms of your hands together as fast as you can. Do you feel something happening to your hands? What you are doing makes heat. The action making the heat is called *friction*. Friction does some other things we can find out about."

SLIP OR STOP

Gather

Slide
Sheet of waxed paper
Rubber sink or tub mat
Sweatshirt
Piece of silk or satin cloth or clothing
Magnifying glass
Small piece of scrap lumber
Small square of coarse sandpaper

Help Find Out

Have your child go down the slide, sitting first on the piece of waxed paper and then on the rubber mat. "Smooth paper doesn't make much friction, so you can move fast. The rubber mat makes more friction, so it's hard to slide." (Although rubber may look smooth, we can feel its stickiness.)

Next, have your child rub an arm first with the sweatshirt and then with the silk or satin. "Which slides faster on your arm? Can you think of a reason why? Look at the sweatshirt and the silk material through the magnifying glass. Which is smooth? Which is fuzzy?" (The smooth-surface fabric slides faster.)

Show your child the scrap lumber and sandpaper. "Feel these. Are they smooth or rough? Look at them with the magnifying glass. Do they look smooth or bumpy? Rub them together hard. Keep rubbing."

Explain that rubbing rough things together makes lots of friction. "Are your fingers feeling warm? Are bits of the wood wearing away? Feel your wood. Is it getting smoother? Is your sandpaper changing? What does friction seem to do to things?" ***Friction produces heat and wears away materials.***

Follow Up

Examine the soles of your shoes. Are some smooth and slippery? Do others have rubber soles? Are sneakers good for running and stopping or do they slip? Are car tires smooth and slippery or more like the soles of sneakers? Talk about slippery roads and sidewalks. How does friction keep us safer?

SMOOTH MOVES

As It Happens

You are adding oil to the crankcase of your car at the service station. Your child wonders why. You offer to help find out at home.

Gather

2 small pieces of coarse sandpaper
Petroleum jelly
Peanut butter
Crackers

Help Find Out

Ask your child to rub the two small pieces of sandpaper together fast. "What is happening to the bits of sand? Look at the sandpaper. Is it bumpy and rough? Would it be smoother if something filled up the bumpy places?"

Have your child spread some petroleum jelly on the paper. Now see if the pieces of sandpaper slide against each other more smoothly. "Is sand rubbing off now?" (No.)

"We put oil in the car motor to help fast moving parts slide past each other more smoothly, so friction won't wear them down or make them too hot." Let your child try rubbing two crackers together. Then use smooth peanut butter to act as a lubricant. When you're finished, this is one science experiment that's good enough to eat!

Follow Up

The next time you have a squeaky door or hinge in the house, show your child a small can of silicone spray or household oil. Mention that the oil or spray makes a smooth sliding surface on objects that rub together. It cuts down friction that makes heat, wears bits away, and makes things squeak.

FRICTION MAKES ITS MARK

As It Happens

While drawing with chalk, your child winces at the scratchy sound. "This chalk needs some oil." You suggest finding out if it will really help.

Gather

Chalk
Petroleum jelly
Sheet of waxed paper
Pencil

Help Find Out

Suggest dipping one end of the chalk into the petroleum jelly. "Now try to draw with that end. Did you solve the scratchy sound problem?" (Yes.) "Can you still draw with it?" (No.) "Do you need friction to draw with chalk?" (Yes.)

"What do you think will happen if you try to draw on waxed paper with a pencil?" (Not much. The waxed coating acts like a lubricant.) "Does friction help you make pencil marks?" (Yes.)

Follow Up

When you encounter skid marks as you're driving on the highway, talk about them as results of friction needed to stop a speeding car in a hurry. Perhaps dark rubber–soled shoes have the same unfortunate effect on your kitchen floor.

*GET A GRIP

As It Happens

Your child is trying to open the bathroom door with wet hands, but doesn't succeed. You suggest drying those hands. Then find out more about friction as a helper.

Gather

Screw-top plastic container
Water
Soap

Help Find Out

Have your child unscrew the top from the jar or container. Then have your child try it again, but this time with wet, soapy hands. "Why is it hard to do now?" (The soap and water act as lubricants.) "What is missing?" (Friction.) Show your child how a towel or a rubber jar-lid gripper helps open stubborn jar lids. "Friction helps us in lots of ways."

Follow Up

At your child's next bath time, explain that a wet bathtub is extra slippery because it is smooth. A rubber bath mat keeps us from slipping because it has more friction. "Friction helps keep us safe from falls in the bathtub. That's why there is a NO RUNNING rule at the swimming pool. Wet feet on smooth tiles can cause hard falls."

From Slide to Glide

As It Happens

Your child is helping unload the grocery cart at the checkout counter and may wonder, "How does the counter move our groceries?" When you get home, you can help find out how rollers help move things like the counter conveyor belt.

Gather

Wide rubber band
3 or 4 cylindrical pens, tinker toy rods, or pieces of dowel
4 or 5 cans of food
Kitchen cutting board

Help Find Out

Use the rubber band and the two cylindrical pens to offer something of the idea of how rollers move a conveyor belt along from underneath the grocery checkout counter. (Use your fingers to keep the rubber band moving around the pens.)

"Now we can find out how rollers help move loads. First, put these cans of food on the cutting board. Shove the board across the floor." (It takes effort to do this.)

Put the pens or rods under the loaded board. "What do you think will happen now when you push the load back and forth? Try it." (It's easy.)

"Can you think what the rollers changed?" (The rollers made less friction, since only three or four narrow surfaces rubbed against the floor.) "Friction slowed down the loaded board. ***Rollers cut down friction to make work easier.***"

Follow Up

Alert your child to rollers that help with all kinds of work: Street sweeper brushes clean up fast. Upright vacuum cleaners have visible rollers underneath to sweep dirt into the bag. A roller moves paper along in the typewriter. Paint rollers spread paint on walls faster than a brush can. Rollers flatten out cookie or pizza dough. Rollers help dishwasher shelves glide forward. Kitchen cabinet drawers may have small white rollers on the underside to cut down friction. Open the cabinet below to get a view of the roller in action as the drawer is pulled open.

Watch giant rollers at work in the automatic car wash. Various kinds of rollers move the escalator, the baggage claim conveyor belt, and the moving walkways at large airports. Rollers keep roller-coasters cars on the track. They turn conveyor belts to move bales of hay up into barn lofts.

BOOKS FOR YOUR CHILD: *The Big Book of How Things Work,* by the editors of *Consumer Guide,* shows the rollers that move escalators and those that move paper through the photocopier. Lincolnwood, IL: Publications International, 1991.

Moving Around/Moving Ahead

As It Happens

Your child finds it hard to steer the shopping cart in the store one day. "This front wheel is stuck!" As you help straighten the wheel, you both notice that the front wheels turn in all directions. "That makes it easier to turn the cart around corners." You offer to find out more about how different kinds of wheels work when you get home.

Gather

Kitchen utility cart or wheeled TV stand on 4 swivel wheels
Toy car with axles exposed or conventional roller skates
(not the roller blade type)
2 matching carry-out beverage lids with slit centers for straws
Plastic drinking straw

Help Find Out

"Let's push this cart around. How does it move?" (It turns easily in all directions.) "Each wheel moves separately. A single wheel attached so it can spin around is called a *swivel wheel*. Front wheels on shopping carts and folding strollers are single swivel wheels." Now push the car (or skate) across the room. "Does it move like the cart?" (No, it only moves straight ahead.) "Turn the car over to see how the wheels are attached. These wheels are joined across into pairs by axles." ***Pairs of wheels work together. Single wheels work alone.***

Have your child roll a beverage lid across the floor on its edge. "Will it keep going straight?" (Sometimes.) "Will it stay upright?" (No.) A single wheel only stays upright and moves straight by itself when it is spinning fast. (Think about the two single wheels on a bicycle.) Now insert the straw in the center slots of the two lids, forming an axle to join the two wheels across in a pair. "How will they roll now?" Wheels joined across in pairs move straight ahead. They stay upright even when they don't move.

Follow Up

At the park have fun using and watching for single wheels that stay upright when they spin fast: bicycles, unicycles, scooters, rollerblades.

At the museum look for single wheels that used to do important work. Some worked upright: waterwheels, paddlewheels for boats, windmills, sharpening stones. Other single wheels like potters' wheels and meal grinding stones lie flat to work.

On car trips watch for impressive pairs of wheels and double pairs of wheels on huge transport trucks. Talk about how things would have to be moved if we didn't have wheels to help.

Gears to Go

As It Happens

Your child has been enjoying a new gear game and later notices the ridges around the bathroom faucet knobs. "Look, faucets are gears, too." You agree that the edges feel like gear edges, "but knobs and gears do different work. Let's look at the work some real gears do."

Gather

Hand-operated eggbeater
Deep mixing bowl
Water
Detergent
Wooden spoon
Gear toys or visible-works clock

Help Find Out

Notches on gear wheels mesh and turn other gears to make something work faster or easier. Examine the beater. "How many gear wheels do you see? The turning knob is attached to a ring gear. Watch what happens to the little gear when the ring gear is turned."

Next, experiment with how the gears make work easier. "Let's try to make soapsuds in this bowl first by beating the water and soap with a spoon. Then try the beater. Which works faster and easier?" Point out that the big gear turns the little gears on the beater blades much faster.

Examine the gear toy or the visible-works clock. Some show how gears can change direction, another important use for the single wheels that are gears.

Follow Up

At mealtime, examine a pepper mill and look for a gear that grinds the pepper.

You and your child can have a wonderful time together taking apart a broken windup clock. Use the tiniest screwdrivers you can find.*

Examine a one-speed bicycle. Turn it upside down and look at the gears. Bicycle gears are called *sprockets*. You can explain that sometimes gears turn each other without touching. Instead, they fit into a chain that moves them. Turn the pedals and watch the chain make the small gear turn. Count the number of times the back wheel turns for each time the pedal goes around once.

BOOKS FOR YOUR CHILD: *Machines and How They Work,* by David Burnie, illustrates beautifully dozens of gears inside clock works on page 10. The gears on ten-speed bicycles are illustrated and explained on page 16. London: Dorling Kindersley, 1991.

RESOURCES: Discovery Toys offers Gearopolis, a construction toy that allows many creative ways to get things turning.

---

*Don't let your child work on the springs; they can uncoil suddenly and cut fingers.

A Groovy Lift

As It Happens

Your child is watching you pull window draperies shut. "Did you know that wheels help me pull the draperies across the window when I pull *down* on this cord? It's another wheel that works alone—a pulley wheel."

Gather

Screw hook
Firm cord, double the length of the floor-to-pulley distance
Small pulley (available at a hardware store)
Bucket filled with blocks
Watch

Prepare

Install the screw hook in a basement joist, or use a ceiling hook for a hanging plant or lamp. Pass the cord over the pulley wheel. Hang the pulley on the screw hook.

Help Find Out

Show your child the pulley. "This pulley is a single wheel with a groove. Pulleys help make lifting and pulling easier. Do you think this pulley could make it easier to lift this bucket of blocks?"

First have your child lift the bucket of blocks with one hand. "Let's see how long you can hold this heavy bucket with one hand. Lift it as high as you can. I'll time you." Record length of time.

Then tie the pulley cord to the bucket. "Now try pulling *down* on this cord to lift the bucket. I'll time you to see how long you can hold it up. You're pulling down on the rope to lift the bucket up. The pulley makes the lifting job easier."

Talk about other ways pulleys help people: They are used on cranes, on sailboats, on flagpoles, on scaffoldings, and inside the casings of older windows. They also help people load big ships, trains, and barns and are used in elevators and escalators. ***Single pulley wheels help people and machines pull down to lift up or pull across.***

Follow Up

Your child can have fun with an easily made pulley hooked to a closet clothes peg or to the top of a toy shelf. All that is needed is an empty thread spool (to act as a grooved wheel) and about eight inches of stiff wire (or use a large, W-shaped paper clip, straightened). Fold the wire in the center. Firmly tuck one end into each end of the spool hole. Hang the wire on the clothes peg or hook and provide string lengths.

BOOKS FOR YOUR CHILD: *What Do People Do All Day?,* by Richard Scarry, has excellent pulley illustrations. New York: Random House, 1968.

RESOURCES: The Mechanics Kit, made by MiniLabs, is a worthwhile purchase. Available in hobby shops, the 30 experiments include pulleys, levers, wheels, and more. Order information from Educational Design, Inc., 47 West 13 Street, New York, NY 10011. Many commercial toys with visible working parts are available: toy clocks, music boxes, and locks. A "look-inside" music box and a pulley-operated cable car kit are both available from HearthSong, A Catalog for Families, P.O. Box B, Sebastopol, CA 95473.

CHAPTER 10

Sound

Sound Watch

As It Happens

A very loud noise startles and frightens your child: the unexpected roar of a truck engine, a jackhammer breaking up concrete for street repairs, a loud clap of thunder. Discovering that sound occurs when something vibrates can help your child overcome the fear of scary sounds.

Gather

Yardstick
Rubber bands
Mug

Help Find Out

Your child can best begin to understand what a vibration is by producing one. Say, "Shake your hands as fast as you can. What are your hands doing?" (Wiggling, wobbling?) "Another word for moving back and forth very fast is *vibrating*. Do your hands look different when they vibrate? Sometimes things move back and forth so fast that they look blurry.

"Listen very hard, now, to find out what else happens.

Let your hands vibrate like that close to your ears. Do you hear a soft, whirring sound? **Sounds are made when something vibrates."**

Place the yardstick on the kitchen counter, extending it one foot from the edge of the counter. Hold it down firmly. Bend the free end down, then release. "What did you see? hear?" Stretch the rubber band over your child's thumbs, held apart. Pluck the rubber band to vibrate it. "What did you see? hear?"

Now, stretch several rubber bands vertically over the coffee mug's opening. Invite your child to pluck them, then ask again, "What did you see? hear? **You saw the vibration, and you heard the sound."**

Sound Sensations

As It Happens
Your child rests a hand on the radio while adjusting the volume. Suddenly the music gets very loud. Your child feels the tingle of vibrations and wonders why.

Gather
Waxed paper
Comb
Windup alarm clock, kitchen timer, or toy music box

Prepare
Cut paper the right size to fit and fold over the teeth of the comb.

Help Find Out
Suggest, "Put your fingers very lightly on the front of your throat and sing a sound like *eeeeeee*. What do you feel? Did something inside your throat vibrate? Yes, you made a sound in there.

"What do you feel when you whisper? Whenever we hear a sound, something is vibrating. Notice what you feel in your lips when you touch them together and hum *mmm-mmmm*."

Show your child the unwound clock, timer, or music box. "Do you feel this vibrating now? Is it making sounds?" Then wind and let your child feel again. "Now what do you feel and hear?" Do this with the radio, TV, or stereo speakers when they are turned off and then on.

"Now try making some music with vibrating air and paper. Hold the paper where it is folded over the bottom of the comb lightly between your lips and hum a tune. The vibrations feel funny, but the sound is nice."

Follow Up

If you have an inexpensive stemmed wine goblet, *create*, *feel*, and *hear* a vibration. While your child holds the goblet stem down on the counter, firmly rub your wet finger around and around the rim until sound is produced. Let your child try. How does the sound change if water is added to the glass? [Different levels of water will produce different tones. A water harmonica uses many tuned glasses of water.] Notice the water vibrating with the glass.

BOOKS FOR YOUR CHILD: *Ben Franklin's Glass Armonica,* by Bryna Stevens, Minneapolis, MN: Carolrohda Books, 1983. This book describes the musical instrument Ben Franklin invented: vibrating tuned glass bowls. Mozart's *Adagio in C Major* and *Quintet for Harmonica, Flute, Oboe, and Cello* were both written for Ben Franklin's glass harmonica.

RESOURCES: Perhaps you have seen the sparkling, costumed performer "Kristalleon" playing water harmonica solos with the Big Apple Circus. His lovely music is available as a CD recording: Vox CD[3] VU9008.

Sound Affects

As It Happens

Noise from a low-flying plane is unbearable. Reacting with covered ears, your child asks, "Why does that hurt my ears?" You offer to help find out.

Gather

Plastic wrap
Metal mixing bowl or a plastic pint-size deli container
Rubber band to fit tightly around the bowl or container
A few cereal flakes
Two feet of plastic hose or a vacuum cleaner hose
Plastic golf club cover tube or plastic vacuum cleaner wand
    (Plastic golf club covers can be purchased inexpensively, for less than one dollar, at most sports stores.)
Cookie sheet
Spoon
Paper towel tube

Prepare

Stretch plastic wrap tightly across the top of the bowl or container. Fasten with the rubber band. Put the cereal flakes on the plastic.

You may want to cover one end of the vacuum cleaner hose and wand with masking tape to make the experiment more hygienic. The covered end will be the "speaking end" of the tube.

Help Find Out

Let your child keep watch on the bowl with the cereal flakes on top while you stand nearby and strike the cookie sheet with the spoon. "What happened? What made the flakes bounce around? Sound vibrations or waves traveled through the air to make the flakes bounce. *We can't see sound waves traveling through the air, but we can see how sound waves affect other things.* Sound vibrations pushed against the air that pushed against the cereal flakes and made them bounce. The jet sound waves pushed hard against the air all around them. They pushed the air inside our ears to make them vibrate hard enough to hurt."

Next, have your child examine the plastic hose or vacuum cleaner hose. "What could be inside this hose?" Put one end next to your child's mouth and the other next to your child's ear. "Whisper your name into the hose. Can you hear yourself? What could be vibrating inside the hose?" (Air.)

Have your child place a hand at one end of the hose and speak into the other end, saying a word like *toot* or *boot.* "Can you feel something pushing on your hand each time you say a word?" (The air in the tube.) *Vibrating air brings sound to our ears.*

Now you and your child can have fun speaking to each other through the long golf club cover tubes or the paper towel tube.

Follow Up

Listen to some music with your child. "Now cover your ears with your hands. Does the music sound different?" Remind your child that one vibrating thing makes the things next to it vibrate. "Air vibrates when something next to it vibrates. That is the way sound is carried by air. It is the way sounds usually come to our ears. The air inside our ears also vibrates, so we hear the sounds."

This is a good time to explain to your child that the inside part of our ears is so delicate that vibrating air can vibrate it. This is how we hear, so we must protect and take good care of our ears.

If your child is frightened by thunder during a storm, give thunder a function. Point out that thunder is vibrating air that can give us information. It can tell us how far away the lightning is that causes the thunder. It takes a certain amount of time for sound to travel from the lightning to our ears. Count the seconds (one–one thousand, two–one thousand, etc.) between the very first flash of lightning and the moment we hear the thunder rumble. It takes five seconds for sound to travel one mile.

BOOKS FOR YOUR CHILD: *Goggles,* by Ezra Jack Keats. In this story, Peter confuses the older boys who are trying to catch him by sending his voice through an empty drain pipe. New York: Collier-Macmillan, 1971.

RESOURCES: HearthSong sells Telefun connecting two wooden telephones with 22 feet of rubber tubing for communication fun. HearthSong, A Catalog for Families, P.O. Box B, Sebastopol, CA 95473.

Solid Sounds

As It Happens

You and your child watch a program about dolphins and the ways they communicate. "But how can dolphins hear noises from each other under the water?"

Gather

Kitchen timer or travel clock
Kitchen shears
Steel, glass, or plastic mixing bowl
Water

Help Find Out

Hold up the ticking timer or clock. "Does this sound loud?" Then place the ticking object on a tabletop. Have your child cover one ear with a hand and put the other ear against the tabletop. "Which way does the ticking sound louder—through the air or through the table?" Listen the same way to your clothes dryer when it is in operation. ***Solid objects carry sound vibrations and make them louder.***

Let your child listen to the sound of the scissor blades opening and shutting. Fill the bowl with water. Have your child press one ear against the filled bowl. "Now I'll do the same thing with the scissor blades in the water. Doesn't it sound different coming through the water? *Water carries sound vibrations and makes them louder.*

Have your child hum a sound and listen. Say, "Now press your ears shut with both hands. Keep on humming. Does it sound the same? different? Which is louder—sound you hear coming to your ears through air, or sound carried by the bony parts of your mouth? Now touch your jawbone while you are humming. Is your jawbone vibrating?" (Yes!)

Follow Up

At the swimming pool, have your child go underwater to find out whether the sound of your voice comes through the water. [Yes, both liquids and solids carry sound better than does air.]

Sounds of String

As It Happens

After you and your child have experimented with the ways that solid objects and liquids carry sound vibrations, you might suggest, "Let's see what *else* will carry sound."

Gather

Light string
Metal spoons
Large darning needle
2 small yogurt cups
Bar of soap
2 buttons

Prepare

Cut a piece of string 2½ inches long for each spoon. Make a hole in the center of each cup with the needle. Cut a piece of string five feet long.

Help Find Out

Hold a short string loosely between your hands. "Do you think loose string will vibrate to make a sound if you pluck it?" Now hold the string very tightly and have your child pluck it again. "Which way makes a sound?" [Loose string vibrates too slowly for audible sound.]

"Now let's add vibrating metal to the string and listen to hear if tight string will carry sound." Tie a spoon to each end of the string. Fold the string in half and hold the folded end to your child's ear. Have your child lean over so that the spoons dangle freely. Swing the string to make the spoons strike each other. "Listen to the sounds traveling up the string to your ear."

With the yogurt cups, make telephones. To stiffen the string ends, pull them across the soap bar. Thread the needle with the string, and pull the string through the holes in the cups from the outside. Attach the string ends to the buttons. "You talk into one cup and I'll listen through the other one. We have to be sure to keep the string straight and tight. Now we know that *tight string can vibrate to carry sound.*"

The Long and Short of Pitch

As It Happens

Your child has been plucking rubber bands slipped over several mugs and cups. "I want to make real music, but these sounds are all alike." You can help your child make higher and lower sounds [pitch] with a cracker box guitar.

Gather

Scissors
Cracker box, 9- or 10-ounce size
Large rubber band
Cork

Prepare

Cut a hole about two inches in diameter near the right end of the box front. Stretch the rubber band lengthwise over the box and the hole, knotting the rubber band if it isn't taut.

Help Find Out

Slip the cork under the rubber band near the left end of the box. Have your child pluck the length of rubber band to the right of the cork. Listen. Have your child continue plucking as you slide the cork slowly toward the hole. "Is the pitch changing? Is it higher or lower? Does the whole rubber band vibrate, or just the side being plucked?" (The cork stops the vibration of the rubber band to its left. Only the part closest to the hole vibrates.)

Help your child notice that *the shorter the vibrating part, the higher the pitch.* (The shorter the vibrating material, the faster the vibration. Shorter and faster vibrations make higher pitched sounds.)

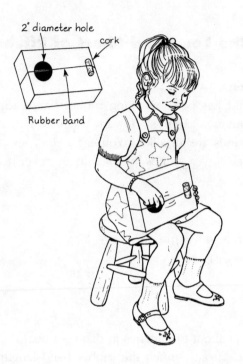

Follow Up

If you have access to an autoharp, you can use this instrument to show the relationship very clearly between string length and pitch. A real harp, an upright piano with the front panel removed, or a grand piano with the top raised will show the same thing. A toy xylophone also shows this relationship.

RESOURCES: A sturdy 15-string Musicmaker autoharp and playing guide for children is available from The Great Kids Company catalog. Call 1-800-533-2166 for a copy.

Tube Tunes

As It Happens

Your child is in the bathtub. You suggest a new way to learn more about the pitch of sounds.

Gather

Funnel

Empty shampoo bottle—a long, slender-necked bottle will
produce a better effect than a cylindrical bottle

Pitcher for water

Help Find Out

In the bathtub, have your child insert the funnel into the shampoo bottle and fill the pitcher with water. Then listen carefully while your child pours a long stream of water from the pitcher into the bottle. Try many times, listening carefully to hear the changing pitch from low to high coming from the bottle. "Is something in the bottle getting shorter and shorter as the bottle fills with water?" [Yes, the air space inside the bottle gets shorter. It's like a tube of vibrating air that gets shorter and shorter.] ***The shorter the vibrating material, the higher the pitch.***

Later, show your child how to make the air space inside the slender-necked bottle vibrate by blowing across the bottle top. Hold lips taut, place lower lip just over the rim, blow across the top with a thrust of air like *tooooooooooo*. Once your child masters the *tooooooooooo* technique, experiment with adding small amounts of water to shorten the vibrating air column, then pouring out small amounts of water to lower the pitch.

Follow Up

All wind and brass instruments in the orchestra depend on vibrating air tubes. Notice the lengths of visible church organ pipes and the chimes in the symphony orchestra.

RESOURCES: Pan pipes show the relationship of length of vibrating air column to pitch. Inexpensive bamboo pan pipes can be ordered from: *Music for Little People* catalog, Box 1460, Redway, CA 95560.

Bouncing Sounds

As It Happens

The child-related noise level in your house is intense. You decide to help your child understand why indoor voices need to be softer than outdoor shouting.

Gather

2 paper towel tubes

Kitchen timer or small travel alarm clock—anything that ticks

Help Find Out

Arrange the two tubes on the counter at a 45-degree angle to each other, with the ends of the tubes facing, but not touching, the wall. Have your child hold one ear next to a tube, holding the other ear covered. You put the ticking timer or clock to the end of the other tube. "What do you hear?" (The ticking sound should reach the wall and bounce back through the listener's tube. The sound is louder when it is confined to the tubes than when it is heard without the tubes.)

Follow Up

Make a stethoscope by inserting a funnel into a 12-inch piece of plastic tubing. Pressed against the chest, the funnel gathers the heartbeat sound so it can be heard through the tube. Place the tubing end close to, but not in, the ear to hear the beats. At your next visit with your child to the doctor, ask to listen through a real stethoscope for a moment.

If possible, experiment with making echoes in an empty gym or some other large, empty space. Explain the echo as the repetition of sound that has bounced off the smooth surface of the opposite wall and returned again to its source.

Listen to the "oceanlike" sounds of a conch shell held close to the ear. Explain that the curves inside the shell pick up sound vibrations nearby. Vibrating air bounces back and forth against the curved walls so much that they sound like the roar of waves along the shore.

BOOKS FOR YOUR CHILD: Ezra Jack Keats' *Apt. 3*. Various sounds guide Sam to a sightless new friend, who knows his neighbors by the sounds he associates with them. New York: Macmillan, 1986.

CHAPTER 11

Light

Getting Comfortable with Darkness

As It Happens

Perhaps your child is almost ready to give up the bedroom night-light or, like many children, is just plain scared of the dark. You can help your child discover that darkness is just the absence of light.

Gather

Scissors
Shoe box with a lid
Tape
Small picture
Flashlight

Prepare

When your child is not around, cut a U-shaped flap near one end of the box lid. Tape the picture inside the box on the end underneath the flap you have just cut. Then cut a dime-sized peephole in the opposite end of the box.

Help Find Out

This works best when the experimenters—you and your child—are together in a semi-dark room at night. Close

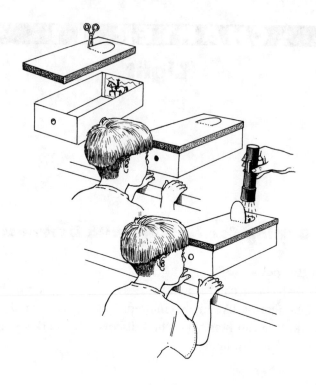

the flap you cut in the lid, but have the flashlight handy so you can easily lift the flap and shine light into the box.

"What do you see when you peek into this special dark box?" Give your child time to think about it. (At this point, nothing is visible.) Then offer to make a change. Push up the flap and shine the flashlight inside the box. The picture can now be seen.

Then say, "The picture was there before, but you couldn't see it. Can you figure out why?" Now let your child control the flashlight to experiment.

"You made the darkness go away when you turned on the light. The picture didn't change. It was always there, but we couldn't see it without light. ***Nothing can be seen without light.***"

Follow Up

Use the concept, **Nothing can be seen without light,** in conversations whenever you and your child step into a dark room or go outdoors on a cloudy night.

Talk about all the furnishings in your child's room with the lights on, then with the lights off, and again to check the facts with the lights on.

You and your child are waiting in the doctor's office, or at the bus stop, or at some other place where boredom can quickly set in. You might want to play an imaginary game about light and darkness. What if . . . all the lights were off in our house for one night. How would you know your own room? the kitchen? the bathroom? the living room? Close your eyes and imagine.

BOOKS FOR YOUR CHILD: *Bedtime for Frances,* by Russell Hoban. Frances discovers, among other things, that the monster she thought she saw in her bedroom when the lights were out was really just her bathrobe draped over a chair. New York: Harper and Row, 1960.

Looking at Light Beams

As It Happens

Your child can't fall asleep and calls from the darkened bedroom, "Why are there stripes of hallway light on my ceiling?" You promise to help find out *another* night, *not now!*

Gather

Pencil
Small flashlight
Empty large cereal box
Scissors
Cardboard

Prepare

Trace a circle around the flashlight lens, in the center of the box front. Cut it out. Cut a matching hole in the box back. Darken the room.

Help Find Out

Hold the flashlight next to one hole. Shine the beam through the holes, toward the wall. Together, notice where the light shines. Does it curve around the box? (No, it shines straight to the wall.) *Light usually travels in a straight line.*

Let your child hold the cardboard in the light beam's path. Does the light beam still go through it to shine on the wall? Is the light shining on both sides of the cardboard? (No, the cardboard blocks the light.) Will your child's hand block the light path in just the same way? Move the hand ahead of the box. (Blocked light leaves a place on the wall where light doesn't shine.) *A shadow is the dark place left when something blocks a path of light.*

Wedge the flashlight through both holes so that the box becomes a holder for the flashlight. Now you can both have fun making shadows in front of the light.

Follow Up

Notice together the straight path of light beams spreading from the projector at the movies, a high-intensity reading lamp in a dim room, car headlights at night, or airport spotlights.

On a very clear night use a strong flashlight beam to point out to your child the stars forming the Big Dipper. (It helps to draw a star-to-star sketch before you go out to stargaze.) Find the North Star at the end of the dipper cup. That star *always* shines in the same place: over the North Pole. (The best place to stargaze is open countryside. In-town stargazing is best away from house or streetlights.)

Look for light-beam motion detectors above automatic doors. When people block the light beams, the blocked light (shadow) signals the doors to open.

BOOKS FOR YOUR CHILD: *The Sky Is Full of Stars*, by Franklin Branley, identifies familiar constellations. New York: Thomas Y. Crowell, 1981.

Changing Shadows

As It Happens

You and your child are walking together outdoors in the noon sunshine. Your child wonders why your shadows are so small. You offer to help find out at home one night.

Gather

Extension cord
Table lamp, preferably spotlight style
Strong flashlight
Small toy figures
Waxed paper

Prepare

Move furniture away from one wall. Use the extension cord to allow the lamp to be moved around the room. Darken the room.

Help Find Out

First, shine the flashlight toward your child's back. Look together at the shadow on the wall ahead. Have your child walk slowly toward the wall, then back away from it. How does the shadow change? Take turns making shadows.

Turn off the flashlight. Put the lighted lamp on the floor at one side of the room. Repeat the shadow walking. Where is the shadow now? How else is it different? Ask, "Where do you think your shadow will be if we put the lamp on the other side of the room?" Explore putting the lamp higher, then adding the flashlight to make two shadows. "You blocked the light path coming from here, so your shadow is there." *A shadow forms away from the light source.*

Later, pretend that the flashlight or spotlight lamp is the

biggest light source of all: the sun. Shine the light on the toy figures in the directions of the morning, noon, and afternoon sun. (Go slowly with this. Avoid suggesting that the sun moves around the earth.) Look at the shadows cast by the figures as the "sunlight" path changes positions. Do the shadows fall in the same direction from each figure?

Now hold a folded piece of waxed paper in front of the light source. Do the shadows change? (Yes, the shadows spread because the light is diffused.) Mention that this is why we don't see our shadows outside on a rainy day. The sun is *always* shining, but the clouds don't let all the sunlight shine through.

Follow Up

If you go to a night ball game played under powerful lights, look at the directions of the players' shadow patterns. Perhaps there is a sundial in your area. Make some sunnyday visits to check the shadow positions, if possible. And when you spend a bright, sunny day outdoors, protect your faces with portable shadows: visors or sun hats!

BOOKS FOR YOUR CHILD: *Dreams,* by Ezra Jack Keats. The giant shadow of a paper mouse saves the night for Archie's cat. New York: Macmillan, 1974. *Shadows Here, There and Everywhere,* by Ron and Nancy Goor. Interesting photographs show how shadows can be useful. New York: Thomas Y. Crowell, 1981.

The Biggest Shadow of All

As It Happens
Your child wonders sadly why it has to get dark at night and bring playtime to an end. You can explore together how night occurs.

Gather
Flashlight
Large rubber ball, earth ball, or globe
Chalk or marker

Help Find Out
Share an important concept indoors in a darkened room. Pretend that the flashlight is the sun that is *always* shining. Pretend that the ball is our Earth. Let your child mark a spot on the ball to represent where you live on the Earth.

Say, "Our Earth is always spinning slowly, even though we can't feel the spinning. What we call one day and night really means that the Earth has made one complete spin around. You shine the sun (flashlight) on the Earth (ball) as I slowly spin it around. Watch the mark showing our part of the Earth. Is the mark in the light now? Then it's daytime on our part of the Earth.

"Watch what happens when our part of the Earth has turned away from the sunlight. Does the sunlight shine through the Earth to light up our side? No, the Earth is blocking the sunlight and making the biggest shadow of all. That shadow is called night! *Nighttime is Earth's shadow.* When we are in Earth's shadow, people living on the other side of the Earth are in daytime sunshine."

Follow Up
Another time, mention that long ago people didn't know what your child knows: that the Earth is a spinning planet

that travels around the sun. They thought the sun rose up in the morning, traveled across the sky, and went down in the evening as it traveled around the Earth! Mention that we still use the words *sunrise* and *sundown* now, even though we know that is only how it looks to us on the spinning Earth.

Experience rotation on a sunny day with a ride on a carousel at the park. "Pretend the ride is the spinning Earth. Now we are in the shadow. We can't *see* the sun, but it's still there. It's almost like nighttime for our side of the Earth and daytime for the other side.

Now we have spun back into the sunlight." The phases of the moon are the shadow cast when Earth blocks the sunlight from shining on the moon. The moon only seems to change its shape.

BOOKS FOR YOUR CHILD: Two books for young children explain the Earth's changing shadow on the moon: Meish Goldish's *Does the Moon Change Shape?*, Milwaukee: Raintree Publishers, 1989; and Franklyn M. Branley's *The Moon Seems to Change*, New York: Harper and Row, 1987. In Franklyn Branley's *What Makes Day and Night*, the Earth's rotation is seen from the vantage point of a space ship. New York: Harper and Row, 1986.

Bouncing Beams

As It Happens

You are driving with your family at night, when headlights of the car behind you reflect from your car's mirrors to your child's face. You ask, "How do you suppose the light behind us can shine in your eyes?"

Gather

Heavy aluminum foil
Flashlight
Small mirror
Ball
Soft pillow

Prepare

Carefully cut six inches of foil from the roll, without wrinkling it. Cut it in half to make two squares. Darken the room.

Help Find Out

"Let's find out what we can see in the room with the flashlight." As your child shines the light on different objects, comment, ***Everything we see reflects some light back.*** Let's see which things reflect light best. Shine the light on your arm. Does your skin look bright and shiny? Shine it on the mirror. What's different now?" (Light reflected back from the smooth, shiny surface of the mirror is almost as bright as the flashlight beam.) Have fun bouncing the reflected light beams around the room.

"Let's see if the smoothness of a surface makes a difference in reflections." Shine the light on the aluminum foil pieces. Look for reflected light on the walls. Keep one piece of foil smooth. Crumple the other piece. Shine the light on each piece in turn. Does the light bounce off each piece just as much? Which surface reflects light better—smooth or wrinkled?

With the lights back on, ask, "Do you think this ball will bounce better off the hard floor or off this soft pillow?" Let your child try both. The soft surface absorbs the force of the ball's energy; the hard surface bounces it back. Soft, irregular surfaces absorb more light energy; hard, smooth surfaces bounce [reflect] the light energy back.

BOOKS FOR YOUR CHILD: *Light and Darkness,* by Franklyn Branley. This book shows the difference between things that create light and those that reflect light. New York: Thomas Y. Crowell, 1975. In Seymour Simon's *Mirror Magic,* children experiment with reflected light. New York: Lothrop, Lee and Shepard, 1980.

RESOURCES: Many kinds of fascinating optical toys, such as periscopes, kaleidoscopes, octascopes, or "optiviewers," use reflected light from mirrors to create interesting effects from multiple images. Look for them in museum shops and better toy stores.

Bending Beams

As It Happens

While peering through an upheld glass of water, your surprised child exclaims, "My fingers look bigger. Why is that?"

Gather

Straight-sided, clear drinking glass
Magnifying lens
Waxed paper
Newspaper
Water
Medicine dropper

Help Find Out

"When we look outside through the flat glass in a window, things look the same as they do when we're outdoors. Do you think things look the same when we see them through a curved glass?" Take turns looking through the drinking glass. [Things look different because the light beams bend as they pass through the curve of the glass.] Feel the curve of the glass in the magnifier. "How do things that are close to you look when you see them through the curved glass of the magnifier?

"Do you think things would look bigger through a drop of water?" Try it. Put a piece of waxed paper on top of some newspaper printing. Carefully release one drop of water from the dropper onto the waxed paper. What does the drop look like? How does the print beneath the drop look now? *Light beams bend when they pass through a curved surface.* So when we look at things through a curved surface, they can appear larger than they actually are.

What happens when your child adds more drops to the

first water drop? Why doesn't the print look bigger through the larger amount of water? (The drop doesn't hold its curved shape. The pool of water is flat now.)

Follow Up

Use a magnifier to examine the curved grooves formed into a plastic card magnifying lens. Keep a card lens in your pocket or purse, ready to explore something interesting with your child.

Look for Dragonfly lenses in toy stores. They create multiple images when you look through the 25 angles (or planes) on their surfaces.

BOOKS FOR YOUR CHILD: *Greg's Microscope,* by Millicent Selsam. Greg examines salt crystals under his microscope. New York: Harper and Row, 1963. *Bending Light,* by Pat Murphy and the Exploratorium. This intriguing book for older children includes directions for making convex lenses of ice or Jell-O! Boston: Little, Brown and Company, 1993.

RESOURCES: Magnifying lenses of many sizes and powers, from tiny magnifier-topped plastic boxes to simple microscopes, binoculars, and telescopes, are available at museum shops and better toy stores. Your child will enjoy taking a closer look at the world with a personal set of magnifiers.

The Secret in Light Beams

As It Happens

Your child is blowing bubbles outdoors and notices the shimmer of colors on the bubbles. "Look, rainbow colors! How did that happen?" You offer to help find out.*

Gather

Clear, rectangular or square plastic box
Water
Prism, cut-glass chandelier drop, or any glass object with corners
Bright sunlight or spotlight lamp
Magazine

Prepare

Fill the plastic box with water.

Help Find Out

"Light beams have a secret inside. Let's see if we can find it!" Have your child peer into a corner edge of the prism, facing toward the sun or lamp light. Encourage patient adjustment until the secret can be seen: the colors of the rainbow [spectrum]! *Light is a mixture of many colors.* Light beams passing at a slant through transparent objects will spread out so the mixtures of colors can be seen.

Use magazine pages as a model to illustrate this concept. Press down on a magazine on a table. Viewed from the side the edges of the pages will look like a solid white

*The earlier experiment, Rainbow Making, on page 69, was designed to explain why rainbows occur; this one helps a child see the spectrum as the "spread-out" colors in light.

line. Now roll the magazine into a tube. The separated pages will fan out and be easily visible. Let your child do this to get a better idea about how "spread-out" [refracted] light beams reveal the bands of colors that make up white light. "We can see the colors in light only when light beams bend to spread out the colors."

Then try looking through the corner angles of other glass objects you may have: a crystal pendant, a carefully handled cut-glass ornament, a water-filled aquarium. Even a tiny breath mint box filled with water will produce a faint spectrum.

Follow Up

When you fill the sink with detergent bubbles to wash dishes, hold handfuls of bubbles up to bright light. Look through the bubbles to see "spread-out" light beams reflected on their surfaces as rainbow colors. Hang a prism in a sunny window. Watch for patches of spectrum colors reflected onto the walls or ceiling.

BOOKS FOR YOUR CHILD: Rose Wyler's *Raindrops and Rainbows* has a simple text that includes experiments about separating light into rainbows. Englewood Cliffs, NJ: Simon and Schuster, 1989.

RESOURCES: Compact discs are mirror-shiny and patterned with dents too tiny to be seen with a magnifier. The patterns break up light into rainbow colors. Certain novelty items seem to make pictures or toy faces appear and disappear when tilted. They use diffraction-grating patterns that break up light into the spectrum. Look for them in toy departments.

CHAPTER 12

Electricity

Charge It

As It Happens

Your child is helping you in the kitchen as you cover a bowl of salad with plastic wrap. Trying to pull out and tear some wrap from the roll, your child protests, "Why won't it stay flat? Why is it sticking to things?" [The experiences with electric charges will not be successful in a humid atmosphere. An indoor, dry winter day would be just right.]

Gather

Plastic wrap
Scissors
Clear, flat plastic box (audio tape box)
Tissue wrapping paper
Combs (plastic or nylon)
Scrap of wool, fur, silk, or nylon
Small pieces of cotton thread

Help Find Out

"You can find out something about why the plastic wrap stuck together when you handle a piece of it differently." Tear off a few inches of wrap and let it lie on the counter briefly. Then cut off a small piece with scissors. "See if this piece will stick to your fingers after you pick it up and let go of it." (It doesn't stick.)

"Now rub the bit of plastic back and forth on your hair, and then try to let go of it." (It clings to the fingers, *unless the hair is damp.* If so, have your child rub the plastic wrap on a silk, wool, or synthetic garment . . . or *your* hair.)

"You just made static electricity. That happens when certain kinds of things rub together. The static electricity caused the clinging when the plastic rubbed against the roll of plastic in the box, too. You can do some fun things making static electricity."

Let your child tear up the tissue paper into rice-size bits. Put some in a flat plastic box. "Now rub the box top with a scrap of fabric very fast. Watch the bits of paper." [They move when charged electrons jump to them.]

After the tissue bits have jumped and clung to the top of the box, turn the box over and repeat the rubbing. (The bits jump to this side.)

"Rubbing (friction) changed some things in the plastic, things so tiny that they are invisible. They are *electrons*. **Tiny, invisible electrons are part of everything in the world—even part of us."** (When some things rub together, part of the electrons hop off one thing and attach to the other. This result is called *static electricity.* The rubbing action is called *charging.*)

"Now try to build a charge on the comb with the fabric. Hold it near the thread. Did the thread pieces jump when the static charge jumped? Hold the charged comb near your hair. Does some hair move with the jumping electrons? *Invisible electrons are part of everything, but we don't notice them until they are charged."*

Follow Up

Watch for rice grains clinging inside a plastic bag as you pour it into a measuring cup. Let your child think about all those rice grains rubbing together, building up static charges inside the bag as they tumble out.

Let your child help with unloading the clothes dryer. Do nylon synthetic garments cling to other things in the load? (They will, unless a fabric softener was used.) The static charge built up when the clothes rubbed together in the revolving dryer. The more the clothes rubbed together, the more static electricity built up.

Engage your restless child in an "unfancy" restaurant by building a static charge on a plastic spoon handle or a plastic pen to attract tiny bits of paper napkin. Try it with salt sprinkled on the dry lid of a drink container. "Will a static charge have the same effect on salt, pepper, sugar?" (Yes.)

BOOKS FOR YOUR CHILD: *Molecules and Atoms,* by Rae Bains, offers a brief explanation of atoms, electrons, and molecules that make up all forms of matter. Published in paperback by Troll Associates, Mahwah, NJ, 1985.

Push Away or Cling

As It Happens
Your child has been having fun using friction to build up charges of static electricity. You offer another way to explore static electricity.

Gather
2 long balloons
2 one-foot pieces of string
Foam plastic meat tray
Wool, silk, or nylon fabric

Prepare
Blow up the balloons. Tie a foot of string to each balloon.

Help Find Out
Ask, "Do you think this tray will stick to a wall by itself? Find out." Then suggest, "Let's see if static electricity will help." Rub fabric very hard on the tray's underside to build up a static charge. "Now will it stick to the wall?" [You built up a charge on the tray, but not the wall. So the negative electrons on the tray jump to the positive electrons on the wall. Unlike charges cling together.]

Next, rub a balloon with fabric to build up a static charge. "Will the balloon stick to the wall?" (Yes!)

Join the two balloons by their strings. Now build a static charge on each balloon. Holding the string, bring the charged sides of the balloons together. "Do they cling together or push each other apart?" [Both balloons built up negative charges. Like charges push apart or repel.]

"There are negative and positive electrical charges. Like charges push apart and unlike charges cling.

Static electricity and magnetism are somewhat alike in this way: The like ends of two magnets also push away from each other and the unlike ends pull together."

Follow Up

Notice how hair flies up toward the comb or brush when you are doing your hair on a dry day. The brush or comb rubs against the hair and makes the hairs push apart. See how certain garments cling to your legs on those days.

*Snap, Crackle, Flash

As It Happens

On a dry winter day your child scuffs across the carpet, touches the doorknob and feels the prick of static electricity. "Ow! What was that?" You offer to help find out at night when static charges can be seen as well as felt. (Your child should wear leather-soled shoes for this experiment.)

Gather

Small, long balloon
Wool or nylon carpet or fabric
Flashlight

Prepare

Blow up the balloon and darken the room.

Help Find Out

"Rub your shoes across the carpet, like you rubbed the box of tissue bits. This time you're building a static charge on yourself! Now that it's dark, watch carefully as you touch the doorknob. What do you see?" (A tiny flash of static electricity is seen as the charge jumps from finger to knob.) Add, "Lightning is a huge jump of static electricity in storm clouds.

"Let's see if we can make a tiny lightning flash. Pretend that our balloon is a thundercloud full of water. Drops of water and drops of ice are blown about and rubbed together in a thundercloud to build up a big charge of electricity."

Show your child how to stroke the balloon on the carpet to build up the electric charge. Count 20 strokes at least. "We need to build up a strong charge."

Switch off the light and use the flashlight to work by. "Now pretend my finger is another cloud or the Earth. I'll bring the charged balloon close to it. Watch! Listen! Did you see the spark jump to my finger? Now you try it. What did you feel?" Let your child experiment.

"The charge made electrons leap across to your finger. The movement made light, sound, and heat. You felt the tiny jolt of heat as a prick."

Follow Up

Turn down the lights near your dryer, or do this in a darkened laundry room at night. Let your child watch as you pull out a load of clothes that has not been treated with fabric softener. Listen and watch for the static sparks as you pull apart the clinging clothes.

Recreate the situation of walking across a rug and touching something metal. But do it in a room that is almost dark. "You can feel, hear, and see the small charge of electricity jump from yourself to the metal. You felt the tiny spark as a little prick. A small spark like this isn't dangerous, but lightning in storm clouds is very powerful. The leap of lightning from storm cloud to Earth is a *powerful* jolt. It can knock people down and hurt them badly. That's why we find a safe place to stay during a bad thunderstorm, like inside the house or car.

"Lightning strikes the tallest thing in its path toward the ground. So we don't stand under the only tree in an open area during a storm. Lightning could hit it and hurt us."

BOOKS FOR YOUR CHILD: Franklyn Branley's *Flash, Crash, Rumble, and Roll* provides excellent information about lightning and thunderstorm safety rules. New York: Harper and Row, 1985.

Connections

As It Happens

The power goes off during a favorite TV program. Your child jokes about scuffling across the rug to make electricity to see the rest of the show. You offer to get some things together later on to find out about the kind of electricity that really runs things: current electricity.

Gather

Wire snips or utility scissors
Buy from a hardware store:
   3 feet of 22-gauge coated bell wire
   6-volt lantern battery
   #502 flashlight bulb with a screw-in base
   Miniature socket to fit the bulb
   Tiny screwdriver

Prepare

Cut two pieces of wire, 12 inches long. Slice through and peel off ½ inch of insulation from each end of the wires with the wire snips. Curl the exposed wire into hooks.

Help Find Out

"When we use an air pump to push air into tires, basketballs, and air mattresses, the pump doesn't make the air. It pushes along the air that comes into it through an air hole. Batteries do something like that, but they don't *make* electricity. *They push the electrons already in something to make them flow along. When electrons keep moving, we call it* **current electricity.** Small batteries like this can push only a small amount of electricity through the wires, so it is safe enough for us to touch. *The electricity that is pumped by giant generators into wires for houses and buildings is so powerful that it is too dangerous for people to touch.*"

"Current electricity can only move in a loop pathway. It moves in a complete loop that we call a **circuit.** We can find out what this means by trying to light this small bulb."

Have your child screw the bulb into the small socket. Then show your child how to loosen a lantern battery terminal cap, hook the end of one wire beneath it, and tighten the cap.

With the screwdriver, loosen, but don't remove, a socket screw. Hook the other end of one wire around it and tighten it with the screwdriver.

Point out, "The light hasn't gone on yet. We need to make a complete loop for the electricity to move on. It has to flow from the dry cell, through the light socket, and back to the dry cell. See if you can use the other wire to complete the loop and make the light go on."

After the light is on, ask your child to find ways to turn the light off. (Unhook or disconnect one end of one wire.)

With a finger, trace the path of the completed circuit. Start at one terminal, follow the wire to the socket. Remove the bulb to show that the metal pathway con-nects to the metal threads of the socket and bulb. Notice that the tiny filament inside the bulb is also a loop. Follow the second wire back to the dry cell. Let your child trace the path.

Follow Up

Examine an extension cord with your child to find the two wires it has side by side. Look at the two prongs on one end and the two holes on the other end.* "Why do you think two wires are needed? Would electricity flow if the extension cord were plugged into itself? What is missing? *You must never play with electrical outlets.* This kind of electricity is powerful enough to hurt people badly if the current goes through them."

BOOKS FOR YOUR CHILD: Herman and Nina Schneider's *Science Fun with a Flashlight* contains a simple description of how electricity from a battery lights a bulb. New York: McGraw-Hill, 1975.

RESOURCES: Melvin Berger's *Switch On, Switch Off* explains how electricity is generated, how it flows to your house, into lights, and back to the generator. It has directions for making a weak current with a bar magnet and wire. New York: Thomas Y. Crowell, 1989.

---

*Many appliances have a third prong for safety. If something causes a short circuit inside the appliance, the third prong will let the electricity flow into the ground safely. That's why the third prong is called a *ground.*

*Carrying Through

As It Happens

You stop your child from reaching for the bathroom light switch with dripping wet hands. You decide that a first hand experience with a simple circuit setup can explain the danger more effectively than words can.

Gather

Use the same 6-volt battery, flashlight bulb, and socket that you did in the previous experiment

Wire snips or utility scissors

Piece of electrical wire 12 inches long

2 pieces of electrical wire 8 inches long

2 roofing nails or large-head tacks

Small block of wood

Test materials such as string, rubber band, sticks, long nail, key, 1-inch tile square, plastic spoon, stone, leather, glass, paper clip, paper, foam packing chip

Prepare

Slice through and peel off ½ inch of insulation from each end of the wire pieces. Hammer the two nails an inch apart into the block of wood. Help your child connect one end of the long wire to a battery terminal and the other to the socket. Then, connect one short wire to the battery and a nail. Next, connect the second short wire to the socket and the other nail.

Help Find Out

Ask, "Why isn't the light shining now?" (The circuit is not complete for the electricity to travel along.)

"Now let's see if we can complete a circle pathway by

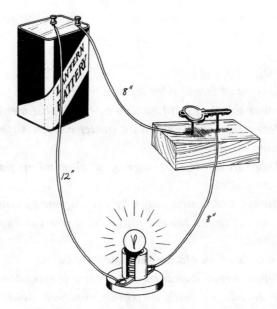

putting a key across the two nails." (The bulb lights. The circuit is complete.)

Let your child try placing other materials across the space between the two nail heads to discover which things let electricity pass and which do not. Emphasize that *some materials conduct electricity; some do not.*

Change one wire, using a new piece with the insulation still covering the end. Does the electricity pass through, even though the circuit *looks* complete? Why not? (The plastic or rubber coverings do not conduct electricity.)

"Water makes some things good conductors of electricity. Wet skin makes our bodies good conductors. Powerful house electricity could pass through a wet person, if wires behind a switch aren't well insulated. The person would be badly burned [electrocuted]. The same thing would happen if an electric appliance should drop into a tub or sink full of water. That is also why lifeguards tell swimmers to come out of the water when an electrical storm starts. If

159

lightning strikes the water near a swimmer, the same thing can happen."

Follow Up

Make a list of family rules to avoid electric shock hazards:

Never touch a light switch or cord with wet hands.

Never put anything in an outlet except safety caps or cord plugs.

Never touch the prongs of a plug when putting it into an outlet.

Always hold cord plugs by the rubber to connect or disconnect them. Do not pull a plug from the socket by the cord.

Never use an old, broken electric cord.

Never, never touch any electric wire outdoors with your hands or with anything you are holding (i.e., rake, kite or balloon strings, or even water spraying from the hose you hold).

BOOKS FOR YOUR CHILD: *Power Up,* by Sandra Markle. Written for older children, this book offers directions for a variety of safe investigations of electricity, as well as good background information. New York: Atheneum, 1989.

CHAPTER 13

Forms and Formulas

Changing Forms

As It Happens

Perhaps your child is fascinated by a snow scene–dome paperweight and wonders why the "snow" doesn't melt into water. You explain that only real snow can change its form into water and that water can even change into another form.

Gather

Ice cube
½ cup microwave-safe ramekin or custard cup
Clear yogurt cup lid to fit the ramekin/cup
Microwave
Magnifier
Freezer space

Help Find Out

It takes just a few minutes to observe three forms of water: solid to liquid to gas; back to liquid; and later, back into sleety solids.

Put the ice cube into the cup, cover with the plastic lid, and pop it into the microwave for one to two minutes. "How do you think it will change?"

Remove the cup with the lid in place. "What do you see *first* as I lift the lid? What do you see next, inside the cup? *Heat changes solid ice to liquid water. Heat changes liquid water to gas (vapor, or steam)."*

Quickly replace the lid and look through it. "What do you see under the lid?" *Water vapor condenses into water droplets as it meets a cooler surface.*

Let the cup cool a bit, with the lid still in place. Examine the droplets under the lid, using the magnifier. Put the cup in the freezer for an hour. Check the contents: solid ice in the cup, ice sleet under the lid where drops of water have frozen. (Snow and frost crystals are frozen water vapor.)

Replace the lid, and let the cup warm at room temperature for ten minutes. Carefully lift the lid. "What do you see?" (Droplets are gathering into larger drops, like raindrops: the whole water cycle in one cup!) *The three forms of matter—solid, liquid, and gas—can be changed under certain conditions.*

Lovely Liquids

As It Happens

Your child is helping with meal preparation and, noticing separated liquids in the salad dressing bottle, says, "Something is wrong with this." You offer to have some fun after dinner finding out about how different kind of liquids act.

Gather

Water

Clear salad oil

White corn syrup or honey

4 plastic deli container lids

2 clear, pint-size plastic soft drink or shampoo bottles with screw caps

Liquid food coloring, red or blue, in a dropper-top bottle

3 thimble-size protective caps from pump-top spray containers (or small, clear plastic vials or bottles)

Help Find Out

Find out that *liquids can differ in the speed with which they flow.* [This is determined by their stickiness, or viscosity.] Pour a small amount of water, oil, and corn syrup or honey on separate plastic lids. Let your child tip the lids to compare the way each liquid moves. "Do they spread and flow the same way?" Touch, smell, and taste, if desired, to learn the characteristics of each liquid.

Fill the two pint bottles within an inch of the top with water. "What do you think will happen when I add a drop of coloring to the water?" (Color will slowly spread through the water until it diffuses completely into the solution.) Let your child add the coloring to the second bottle and quietly watch the diffusion.

Let your child slowly pour some syrup or honey into one of the bottles of tinted water. "What's happening?" (Syrup settles to the bottom before diffusing.) [The more viscous liquid is heavier.] "Do you think the same thing will happen if we add oil to the other bottle?" *(Oil is lighter than water,* so it floats on top.) Find out what happens when the tightly capped bottles are shaken vigorously. (Syrup dissolves. Tiny oil droplets scatter into the water.) Let stand for a few minutes. "Do the oil bits stay scattered in the water?" (No, oil droplets collect together again. *Oil does not dissolve in water.*)

Fill the three clear caps or vials to the depth of one inch each with water, oil, or syrup. While your child watches, at eye level with the container, carefully add one drop of food coloring to each liquid. (Drops will diffuse quickly in water, spread across the top of the syrup, and stay suspended like small jewels in the oil.) Later, gently stir the suspended coloring with a straw. (Coloring will diffuse into a solution in the syrup, but break up into fine droplets as an emulsion in the oil.)

Follow Up

Watch the air bubble sealed into the spangles and liquid-filled plastic magic wands available at toy and novelty stores.

Speculate with your child about the two liquids in a spiral drips timer or those in the colored "wave pictures" found in museum or novelty shops. Talk about the viscosity of various liquids on the table at mealtime: sauces, dressings, beverages. Which one would spread fastest if spilled? [Let's try *not* to find out at the table!]

It Shrinks, It Stretches,
It's a Gas!

As It Happens

You tie a bunch of birthday party balloons to a sunny front porch. Later, the biggest balloon pops. Your child wonders, "How could that happen?" You promise to help find out after the party.

Gather

Uninflated small balloon
Small-necked plastic bottle, such as a 10-ounce spring water
   bottle
Hottest water from the faucet
Small saucepan

Help Find Out

"This balloon can help us find out what will happen when we change the temperature of the air in the bottle."

Stretch the neck of the balloon over the plastic bottle top. Place the bottle in the freezer for half an hour. "What has happened to the balloon?" (It will be slightly sucked into the bottle.) *"Cool air contracts.* The air pulled closer together when it got so cold."

Quickly, put the bottle in the pan and hold it under the stream of running hot water. "What might change as the air inside gets warmer?" (The balloon will suddenly expand to stand upright. *Warm air expands.*)

"Air spreads out and gets lighter when it is warmed. That's what happened when the sun warmed the air inside the birthday balloon. When the balloon couldn't stretch anymore to hold the spreading air, it popped!"

Explain that *air is a mixture of gasses. A gas doesn't*

have its own shape. It can be squeezed into a small space, or it can expand and fill more space when it is warmed.

Follow Up

The next time you have an empty thin plastic jar or bottle to recycle (it will have a recycling symbol on the bottom), do the reverse of this experiment as you and your child work together in the kitchen. Fill the container with the hottest tap water. (Avoid boiling water, as it will melt thin-gauge plastic.) Leave the hot water in the bottle, uncapped, for one minute. Empty it and immediately screw on the cap. Leave the container on the counter. Let your child watch it, while you go on with your work. Eventually, the container will collapse with a *snap!* Help your child think about why. (Air inside cooled and shrank. It didn't press as much on the bottle, at that point, as did the outside air. The outside air pressure made the bottle collapse.)

BOOKS FOR YOUR CHILD: Two fine books are plotted around hot-air ballooning: *Hot-Air Henry,* by Mary Calhoun, New York: William Morrow, 1981; and *The Great Town and Country Bicycle Balloon Chase,* by Barbara Douglass, New York: Lothrop, Lee and Shepard, 1984. Both are beautifully illustrated to show the fans that inflate the balloons and the burners that warm the air to lift the balloons.

Busy Bubbles

As It Happens

After your child has become aware of gas as a form of matter, this experience can make a slow day more interesting.

Gather

Screw-top bottle of colorless soda
Small handful of chocolate chips, raisins, grapes, or peanuts
Clear plastic tumbler

Help Find Out

"Watch and listen carefully while I unscrew the cap on this soda bottle. What did you notice?" (The hiss of escaping gas, and gas bubbling to the top of the bottle.) Pour the soda into the tumbler. Discuss what is moving to the surface of the liquid. (Carbon dioxide gas was dissolved into the liquid at the bottling factory.) Decide if the gas is lighter than the liquid. "Is that why the bubbles of gas travel up to the top instead of going down?" (Yes.)

Let your child drop the chips, nuts, or fruit into the tumbler of soda and enjoy watching closely as tiny bubbles gather on the surface of the chips and make them buoyant enough to ride to the surface. What happens to the chips as the bubbles of gas escape into the air? (Without enough clinging bubbles, the chips sink, until fresh bubbles collect to take them up again.)

Your child can see how long there are enough escaping gas bubbles clinging to the chips and pushing them to the surface. It might go on for several hours. When the chips finally come to rest at the bottom of the tumbler, the experiment can be eaten. Does the soda now seem uninteresting? What is missing?

A sick-in-bed child might like to compare two or three different nonsoluble edibles to see if the activity varies in some way. Set up as many small glasses of soda as the bedside table can hold. An older child can make a chart to time the various results.

Staying Solid

As It Happens

You have pulled from the bottom of the dishwasher a crumpled glob of plastic that once was a plastic spoon or container lid. Show your child that the dish-drying temperature changed the shape but that the glob is still a solid bit of plastic. It didn't become different stuff. Then have fun changing the forms, but not the makeup, of some solids.

Gather

Microwave
Chocolate chips
Small microwavable cup
Cornstarch
Water
Measuring cup and spoon
Plastic container lid or saucer

Help Find Out

Briefly microwave some chocolate chips in the cup. Let your child describe the thick, liquid chocolate. Then do a taste comparison of the melted liquid and some unmelted, solid chips. Even though it changed form, is it still chocolate? (Yes, it didn't mix with anything to become something different.)

Let your child stir and knead together ¼ cup of cornstarch with two tablespoons of water in the saucer to make a semi-solid that looks and acts somewhat like a liquid as it spreads out. When your child tires of playing with this "goop," allow it to dry overnight as a thin layer. Check it the next morning to see that it is still cornstarch that can be crushed back to its powdered form. (The water changed the form of the corn-starch temporarily, but the water evaporated into the air overnight, leaving the solid cornstarch.)

Follow Up

Point out other solids that can change their form but not their makeup when warmed: butter or ice cream, for example.

*Making Mixtures

As It Happens

You are assembling ingredients to make cookies. Your child surreptitiously tastes the baking soda and warns, "Don't put that stuff in the cookies. They will taste terrible!" While the cookies are baking, you help find out how soda changes the other ingredients into cookies.

Gather

½ teaspoon baking soda
4 teaspoons water
2 teaspoons brown sugar
2 microwavable saucers
Microwave oven

Help Find Out

Stir two teaspoons water, one teaspoon brown sugar, and the ½ teaspoon baking soda together to dissolve in one saucer. Microwave for one minute, watching closely to see the chemical reaction. (Heat is needed for this reaction.)

Carefully remove the saucer when bubbles begin to get large. (The mixture starts to burn quickly after this stage.) *Do not touch* the hot mixture. It's burning hot and sticky. "How has the mixture changed?" Let it cool in a safe place.

Repeat the experiment in the second saucer, using the other teaspoon of brown sugar and two teaspoons water, without adding baking soda. Compare the results. (Heat brought about a chemical reaction between the soda, water, and brown sugar, but not between the sugar and water alone.) "This first mixture formed bubbles of gas and changed the ingredients into something different. ***When chemicals mix and react, they change into something***

different. They don't stay the same. The soda in the cookies also didn't stay the same in the oven's heat. It reacted with other ingredients to change into gas bubbles just like the ones in the saucer. The cookies got bigger and lighter. The soda didn't stay the same, so it doesn't make the cookies taste bad. Try one and see for yourself!"

Follow Up

Read the list of ingredients on all the cracker and cookie boxes on the shelf to find out which ones use baking soda to make them light and crisp. Examine the inside of a broken cracker through a magnifying glass to see the air spaces created by the gas bubbles.

BOOKS FOR YOUR CHILD: *Messing Around with Baking Chemistry,* by Bernie Zubrowski, provides further information for older children about kitchen chemistry, experiments, and recipes. Boston: Little, Brown and Company, 1981.

CONSUMER ALERT: Avoid purchasing "kitchen chemistry" sets. The contents can be disappointingly skimpy or stale; the suggested activities are often poorly organized and explained.

Expansive Reactions

As It Happens

Your child is puzzled to learn there is baking soda in your new toothpaste. "Why? We don't bake with toothpaste!" You can suggest finding out that soda can do many different things by trying this chemical surprise.

Gather

¼ cup plain white vinegar (5 percent acidity)
Plastic pint jar
1 tablespoon baking soda
Plastic sandwich-size bag (not the zip-top or flap style)
Rubber band

Help Find Out

"What do you think might happen if you mix these two chemicals together: baking soda and vinegar?" [Chemically speaking, vinegar is an acid; the baking soda acts as a base.]

Let your child pour the vinegar into the jar. You put the baking soda into the bottom of the plastic bag and, letting the bottom of the bag hang down outside the jar, fasten the open end of the bag over the mouth of the jar with the rubber band.

Let your child lift the bag to tip the soda into the jar. "Watch the reaction as the chemicals mix together." (The result will be foaming bubbles and an upright bag full of carbon dioxide gas!) Leave the bag of captured gas untouched for as many days as your child wants to keep it. If your child wants to try the experiment again, be sure to start with a clean jar. Ask for a prediction of what will happen. It's possible to predict the result because *when the same chemicals are mixed together in the same amounts, the reaction will be the same.*

Follow Up

Emphasize to your child that some chemicals are too dangerous for children to use. Only adults who know how to use
them can do so safely. But baking soda and vinegar are safe
enough chemicals to put into our food, so children can safely
have fun making a sandbox volcano in the following way:

Stir together two tablespoons of baking soda, a few
squirts of liquid detergent, and a half cup of warm water.
Pour the mixture into a clean, narrow-necked plastic pint
bottle. At the sandbox, surround the bottle with a "moun-
tain" of damp sand. Pour ¼ cup of white vinegar into the
bottle. Voila! A foaming volcano. You may need to try com-
bining the ingredients in varying amounts, since acidity levels and bottle shapes will affect the reaction.

To explain the Fourth of July fireworks explosions or spaceship blast-off launchings, consider doing the Expansive Reactions experiment carefully outdoors, quickly sealing the bottle with a wet cork after adding the water and vinegar. To avoid possible injury, aim the cork *away* from everyone, since the reaction can launch the cork forcefully. The explosion of fireworks happens when stronger chemicals make a great deal of gas very quickly, resulting in the bang and blast-off.

Another time, let your child work independently to find out which substances have acid content. Spoon a small amount of baking soda onto several deli container lids. Provide small amounts of "test materials" in discarded film containers, old prescription vials, or other small plastic bottles. Try: orange, lemon, or apple juice; buttermilk; regular milk; 7-Up; the separated liquid from a jar of mustard; pickle juice; liquid soap. The lids need to be washed if used for more than one test material. An older child will enjoy keeping a YES/NO chart to record acid materials that foamed into gas bubbles with the baking soda. Which substances foamed a lot? (The stronger, more acidic materials.) Which foamed only a little? (The milder, less acidic materials.)

CONSUMER ALERT: In the interest of child safety, here is a book to censor or *avoid* reading to your child: a 1989 adaptation of *The Peterkin Papers,* written in 1867 by Lucretia P. Hale—*The Lady Who Put Salt in Her Coffee.* This charmingly illustrated book could be dangerous to your child's health, since it records many poisonous chemicals being sampled in a cup of coffee. Surprisingly, the book jacket bears the seal of a Parents' Choice Award.

Show to Tell

As It Happens
Your child enjoyed experimenting with baking soda reactions and is ready for new adventures in kitchen chemistry.

Gather
Advance preparation:
Fresh red cabbage
2-cup measure or small glass or steel bowl
Boiling water
Sieve
Clean pint jar

When you are ready for the experiment:
2 clean baby food jars
Measuring spoons
White vinegar (ordinary 5 percent acidity)
Baking soda
Plastic medicine dropper*

Help Find Out
Coarsely chop enough red cabbage to fill the two-cup measure or bowl. Add boiling water to cover cabbage. Let stand for one hour. Pour resulting purple liquid through the sieve into the jar. (This liquid may stain your counter, so wipe spills immediately.) You have made a natural *indicator* solution. "This beautiful liquid will tell us if something is an acid or a base."

Do this activity at the sink to minimize spill problems.

---

*You can improvise a substitute medicine dropper by folding a plastic straw almost in half and squeezing the two sections together to draw up and dispense liquids.

Pour ½ cup of the indicator solution into each small jar. "This solution isn't an acid or a base, but it reacts to acids and bases in surprising ways." Let your child add one tablespoon of vinegar to one jar of solution to see what happens. (It changes to reddish pink, indicating that vinegar made the solution acidic.)

"What do you think might happen when you put a teaspoon of baking soda in the other jar of indicator solution? Find out." (It turns blue, indicating a mild alkaline, or base, solution.)

"What do you think might happen if you add vinegar to the blue [alkaline] liquid? Try adding a bit at a time with the dropper." (About two teaspoons of the white vinegar changes the solution back to a purple, or neutral, indicator solution.)

"Now try adding baking soda to the pink [acidic] liquid. What do you think might happen?" (About one teaspoon of baking soda results in a purple indicator solution.) "It's as if acids and bases are on different sides of a teeter-totter. Too much on the acid side makes it pink. Too much on the base side makes it blue. Just the right amount of each balances it back to purple again. *Acids and bases are chemical opposites. They can neutralize each other."*

Offer your child time to experiment alone with small amounts of indicator in *clean,* clear spray bottle protector caps and small plastic containers, such as aspirin bottles. Test for the acidity of lemon juice and buttermilk; the alkalinity [a base dissolved in water becomes an alkali] of liquid soap and milk of magnesia. Try to find other mild acids or alkalis to test. An older child may want to keep a crayoned record of the indicator colors: the stronger the acid or alkali, the darker the color.

Recall with your child the safey cautions about experimenting independently. You might want to demonstrate to your child why the dishwasher detergent box has a warning label about careful use. Add ½ teaspoon of it to ¼ cup of indicator solution to see a dramatic change to dark green. Strong alkaline solutions can burn the skin.

Follow Up

When you and your child see ads for antacids to neutralize stomach acids, recall what that means. Try dissolving such a tablet in a weak acid solution to verify what happens. Add that "our wonderful bodies usually make just the right chemicals from the food we eat."

Mention that bee stings are acidic, so putting baking soda on the sting neutralizes the acid and helps the sting feel better.

The next time you brew regular tea for yourself, pour an extra-strong amount of tea in a white or clear glass cup. Let your child find out that tea is another natural indicator by noticing what happens when a teaspoon of lemon juice is added to the tea, ¼ teaspoon at a time. Compare the color to that of your cup of tea. Ask for a prediction of what might happen when ½ teaspoon of soda is added a bit at a time. (The tea foams and its color is restored.)

If you keep an aquarium, let your child share in checking the results of water pH testing. Explain pH as just the right amount of base in the water to keep the fish healthy.

*Fire Facts

As It Happens

Perhaps your child has watched you fan a fire to start it burning faster and wondered if you were trying to blow out the fire. You can offer to help find out the three things a fire must have to burn. This information will make the fire safety rules understandable.*

Gather

Matches
Votive candle in holder
Paper towel
Small pitcher of water
6-inch candle in holder
Empty pint jar
Carbonated soft drink
Mug
Kitchen scissors

Help Find Out

Begin with the safety rule: "ONLY ADULTS STRIKE MATCHES. CHILDREN NEVER EXPERIMENT ALONE WITH FIRE. IT'S TOO DANGEROUS. *It takes three things to make a fire burn: heat, oxygen, and fuel. Fire is a chemical reaction among those three things.* Let's find out what happens to a candle flame when we take away each one of those things. Remember, you mostly watch these experiments."

Light the votive candle. Watch the wax melt as the burning wick heats it. "This is the fuel that lets the candle wick burn."

*Thanks go to the Racine, Wisconsin, firefighters who helped clarify these experiments.

Let your child hold the end of a thin strip of paper towel in the pitcher of water. Watch the wicking action as the water spreads up the strip. "The melting candle wax soaks into the candle wick in the same way. Candle wax makes good fuel."

Point out the flame as the *heat* at the kindling point, the wax as the *fuel,* and the air around the candle as the *oxygen* source.

"Can you think of a way to *push* the heat away from the candle fuel?" (By blowing it away with a quick, strong puff!) "This works for small candle flames, but *not* for big fires. Fast-moving air feeds big flames more oxygen, so the flames burn fuel faster."

Suggest, "Let's see what happens to the candle flame if we take oxygen away from it." Invert the jar over it to limit

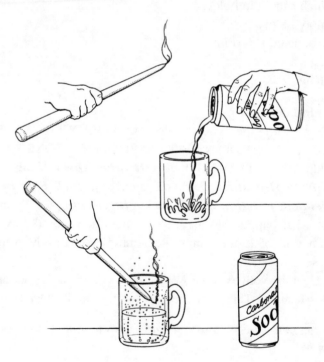

the supply of air to the flame. "Watch what happens." (The flame soon stops burning.) Recall that oxygen is part of the air in the jar. "The fire went out after it used up all the oxygen in the jar."

If you have done the Expansive Reactions experiment on page 173 of this chapter, show another way to take away the oxygen to put out the flame. Light the long candle. Then open a fresh can of carbonated drink and quickly pour the mug half-full. Immediately lower the candle flame to just above the liquid.

As you watch the flame go out, recall the way carbon dioxide gas filled the plastic bag in the Expansive Reactions experiment. "The carbon dioxide from the drink is beginning to fill the cup. It has pushed air away, so the burning wick couldn't get enough oxygen. That's why the flame went out." Add that taking the oxygen away from a fire with carbon dioxide or other chemicals is one way that firemen put out some fires. "That's why the STOP, DROP, and ROLL rule works to put out fires in clothing. It cuts off the oxygen. Running feeds more oxygen and makes the flames burn faster."

Suggest, "Now let's try to take the fuel away from the flame." Light the votive candle and let it burn long enough to form a full pool of melted wax at the base of the wick. "What do you think will happen if we cover the melted wax with water?" Drip a spoonful or so of water onto the pool of wax. [Water covers the fuel and soaks into the wick, extinguishing the candle.] "Water isn't fuel. It can't burn, and that's why it is used to put out many kinds of fires."

"Let's see if we can find another way to take the fuel away from the flame." Light the long candle again. With the kitchen shears, carefully pinch off and hold on to a small bit of the burning wick. (Do this over the sink, so that nothing is damaged and the experiment isn't lost if the bit of burning wick falls from the scissors.) "Do you think this bit of wick will keep on burning?" (It goes out when the fuel is used up and no more wax is available as fuel.)

Follow Up

If you have a candle snuffer, let your child use it after a festive meal to deprive burning candles of oxygen, putting out the flames.

If you are grilling outdoors or building a fire in the fireplace, talk about keeping other flammable materials at a safe distance. "We only want the charcoal (or wood) to be the fuel for this fire; nothing else."

When you cook with your child, point out that paper, cloth, plastics, and other flammables placed on top of a cooking or heating unit will become fuel and catch fire from the heat.

BOOKS FOR YOUR CHILD: *A Chair for My Mother,* by Vera Allen, illustrates the aftermath of a house fire and how a loving family recovers from it. New York: Greenwillow, 1984.

RESOURCES: Learn about community resources for fire safety instruction to reinforce these important lessons. Send for fire safety tip pamphlets for home and family to National Fire Protection Association, Batterymarch Park, Quincy, MA 02269.

Shiny Reactions

As It Happens

You are getting ready to set a festive table for a family gathering and need to polish the silverware in a hurry. You invite your child to watch some special chemistry help with the task.

Gather

For the first experiment:

Saucepan or skillet (*not* aluminum)

Aluminum foil

Silverware (not recommended for hollow-handled knives)

Water

Salt

Baking soda

Stove

Dish detergent

For the second experiment:

¼ cup distilled white vinegar (the cheapest kind)

2 teaspoons salt

Small plastic cup

Dingy pennies

Old toothbrush

Bowl of water

Paper towels

Help Find Out

For the first experiment in cleaning silverware, show your child the tarnish you want to remove. Set a tarnished piece of silverware aside for comparison.

Cover the bottom of the pan with the foil. Lay the silverware on the foil with each piece touching the foil. Cover the silverware with water, adding one teaspoon of salt and one teaspoon of baking soda. Boil for a few minutes.

Carefully lift out the silverware and wash with regular detergent. Compare the shining silver with the tarnished piece you set aside earlier. Where did the tarnish go? [Tarnish is sulphur. It left the silver and combined with the aluminum.] Show your child the darkened piece of aluminum. [A more complicated interaction, electrolysis, brought about the reaction with these simple chemicals.] Use this method infrequently since overuse can be hard on some kinds of silver plating.

For the second experiment, mix the vinegar and salt in the cup. Let your child scrub each penny with the toothbrush dipped in the solution. Rinse pennies in water; polish dry with the paper towels. (If your child feels very ambitious, the same solution can be used to shine your brass candleholders and copper kitchenware. These need to be polished and scrubbed with a soft cloth.)

Follow Up

Discover how shiny hair becomes when vinegar diluted with water is used as a rinse to neutralize the last traces of shampoo. Mention the chemical reaction.

Even a young child will take pride in polishing hard-water spots off chrome tub faucets with a dab of white vinegar on a paper towel. The interaction between vinegar and the salts in the water spots make this quick work.

Expand your child's participation in home tasks by replacing toxic cleaning products with safe to use, effective, and inexpensive homemade mixtures. Here are two helpful chemical reactions: A mixture of ½ cup baking soda and two tablespoons white vinegar makes a good tub-scrubbing compound. Your child can mix into a paste two tablespoons flour, ½ teaspoon salt, and two tablespoons white vinegar to use as a brass or copper polish. The paste is rubbed on the item, which is allowed to stand 30 minutes, rinsed, and polished with a soft cloth that has been dampened with vegetable oil.

RESOURCES: If you want to make your home a "pollution-free zone," formulas for many other inexpensive, safe alternatives to toxic cleaning products can be found in *Clean and Green: The Complete Guide to Nontoxic and Environmentally Safe Housekeeping,* by Annie Berthold-Bond, Woodstock, NY: Ceres Press, 1990; and *Rodale's Book of Practical Formulas,* Rodale Press, 1991.

APPENDIX
Children's Science Museums

There are currently more than 400 children's museums in the United States and Canada. The list below includes only those that offer hands-on, interactive exhibits in the physical sciences.

Alabama

Discovery Place of Birmingham
1320 22nd Street South
Birmingham, AL 35205

Arizona

Arizona Museum of Science and
  Technology
80 North Second Street
Phoenix, AZ 85004

Flandrau Science Center
University of Arizona
Tucson, AZ 85721

California

Exploratorium
3601 Lyon Street
San Francisco, CA 94123

Lawrence Hall of Science
University of California
Berkeley, CA 94720

Lawrence Livermore National
  Laboratory
University of California
Visitors Center/Tour Programs
Livermore, CA 94551

Reuben H. Fleet Space Theater &
  Science Center
P.O. Box 33303/Balboa Park
San Diego, CA 92163

Florida

Miami Museum of Science and
  Space Transit Planetarium
3280 South Miami Avenue
Miami, FL 33129

Museum of Science and History
1025 Gulf Life Drive
Jacksonville, FL 32207

Museum of Science and Industry
4801 East Fowler Avenue
Tampa, FL 33617

Orlando Science Center
810 East Rollins Street
Orlando, FL 32803

Georgia

National Science Center
Building 25722
Fort Gordon, GA 30905

SciTrek
395 Piedmont Avenue, Northeast
Atlanta, GA 30308

Idaho

The Discovery Center of Idaho
131 Myrtle Street
P.O. Box 192
Boise, ID 83701

Illinois

Chicago Academy of Sciences
2001 N. Clark Street
Chicago, IL 60618

Discovery Center
711 North Main Street
Rockford, IL 61103

Lakeview Museum of Arts &
  Sciences
1125 West Lake Avenue
Peoria, IL 61614

Museum of Science and Industry
57th Street and Lake Shore Drive
Chicago, IL 60637

SciTech: Science & Technology
  Interactive Center
18 West Benton
Aurora, IL 61614

Indiana

The Children's Museum of
  Indianapolis
P.O. Box 3000
Indianapolis, IN 46206

Children's Science & Technology
  Museum
523 Wabash Avenue
Terre Haute, IN 47403

Iowa

The Science Center of Iowa
4500 Grand Avenue
Des Moines, IA 50312

Science Station
427 First Street, Southeast
Cedar Rapids, IA 52401

Kentucky

Museum of History and Science
727 West Main Street
Louisville, KY 40202

Maryland

Maryland Science Center
601 Light Street
Baltimore, MD 21230

Massachusetts

Boston Children's Museum
300 Congress Street
Boston, MA 02114

Museum of Science
Science Park
Boston, MA 02114

Springfield Science Museum
236 State Street
Springfield, MA 01103

Michigan

Ann Arbor Hands-On Museum
219 East Huron
Ann Arbor, MI 48104

Children's Museum
432 North Saginaw
Flint, MI 48502

Detroit Science Center
5020 John R. Street
Detroit, MI 48202

Hall of Ideas
Midland Center for the Arts
1801 West Saint Andrews
Midland, MI 48640

Impression 5 Science Museum
200 Museum Drive
Lansing, MI 48933

Minnesota

The Science Museum of
Minnesota
30 East 10th Street
St. Paul, MN 55101

Nevada

Lied Discovery Children's
Museum
833 Las Vegas Boulevard North
Las Vegas, NV 89101

New Hampshire

Science Enrichment Encounters
324 Commercial Street
Manchester, NH 03101

New Jersey

Liberty Science Center
75 Montgomery Street
Jersey City, NJ 07302

New York

Buffalo Museum of Science
1020 Humboldt Parkway
Buffalo, NY 14211

The Discovery Center of Science
and Technology
321 South Clinton Street
Syracuse, NY 13202

New York Hall of Science
47-01 111th Street
Flushing Meadows–Corona Park,
NY 11368

The Schenectady Museum and
Planetarium
Nott Terrace Heights
Schenectady, NY 12308

The National Soaring Museum
Harris Hill, R.D. #3
Elvira, NY 14906

Sciencenter
601 First Street
Ithaca, NY 14850

North Carolina

Catawba Science Center
243 Third Avenue, Northeast
P.O. Box 2431
Hickory, NC 28603

Nature Science Center
Museum Drive
Winston-Salem, NC 27105

North Carolina Museum of Life
& Science
433 Murray Avenue
P.O. Box 15190
Durham, NC 27704

Science Museums of Charlotte,
Inc.
301 North Tryon Street
Charlotte, NC 28202

Ohio

Cleveland Children's Museum
10730 Euclid Avenue
Cleveland, OH 44106

COSI
Ohio's Center of Science and
Industry
280 East Broad Street
Columbus, OH 43215

Oklahoma

Omniplex Science Museum
2100 Northeast 52nd Street
Oklahoma City, OK 73111

Oregon

Oregon Museum of Science &
Industry
1945 Southeast Water Avenue
Portland, OR 97214

Willamette Science & Technology
Center
2300 Leo Harris Parkway
Eugene, OR 97401

Pennsylvania

The Carnegie Science Center
Allegheny Square
Pittsburgh, PA 15212

The Franklin Institute Science
Museum
Benjamin Franklin Parkway at
20th Street
Philadelphia, PA 19103

The Museum of Scientific
  Discovery
Strawberry Square; Third and
Walnut Streets
P.O. Box 934
Harrisburg, PA 17108

South Carolina

South Carolina State Museum
P.O. Box 100107
Columbia, SC 29202

South Dakota

South Dakota Discovery Center
  & Aquarium
805 West Sioux Avenue
Pierre, SD 57501

Tennessee

American Museum of Science &
  Energy
300 Tulane Avenue
Oak Ridge, TN 37830

TVA Energy Center
1101 Market Street MR 2C
Chattanooga, TN 37402

Texas

The Children's Museum of
  Houston
3201 Allen Parkway
Houston, TX 77019

Don Harrington Discovery Center
1200 Streit Drive
Amarillo, TX 79106

Fort Worth Museum of Science
  and History
1501 Montgomery Street
Forth Worth, TX 76107

Science Spectrum
5025 "J" 50th Street
Lubbock, TX 79414

The Science Place
1318 Second Avenue
Fair Park
Dallas, TX 75210

Utah

Hansen Planetarium
15 South State Street
Salt Lake City, UT 84111

Vermont

Montshire Museum of Science
P.O. Box 770
Norwich, VT 05055

Virginia

Commonwealth of Virginia
Science Museum of Virginia
2500 West Broad Street
Richmond, VA 23220

Science Museum of Western
  Virginia
One Market Square
Roanoke, VA 24011

The Virginia Discovery Museum
524 East Main Street
Charlottesville, VA 24011

Washington, D.C.

National Air and Space Museum
Smithsonian Institution
Washington, D.C. 20560

National Museum of American
  History
Smithsonian Institution
Washington, D.C. 20560

Canada

Saskatchewan Science Centre
Winnepeg Street and Wascana
  Drive
Regina, Saskatchewan
CANADA S4P 3M3

Science World British Columbia
1455 Quebec Street
Vancouver, BC
CANADA V6A 3Z7

Wisconsin

Discovery World
Museum of Science, Economics,
  and Technology
818 West Wisconsin Avenue
Milwaukee, WI 53233

About the Authors

Jean Harlan is a practicing clinical psychologist who has been writing and teaching about child development, parenting, and science education for more than twenty years. She is the author of two textbooks, *Science in the Kindergarten* and *Science Experiences for the Early Childhood Years*, and speaks to professional groups throughout the country. She has served as a science consultant for *Sesame Street,* and was an editorial board member for the National Association for the Education of Young Children. Harlan lives in Oak Creek, Wisconsin.

Carolyn Quattrocchi is a freelance writer and editor on early childhood education and has written nearly a dozen children's books. She also created and edited *My Own Magazine*, devoted to children ages two to six. She previously served as director of Ohio University's Child Development Center. Quattrocchi lives in Evanston, Illinois.